AUSTRALIAN LORIKEETS

DEDICATION

I would like to dedicate this book to
my mother Gladys and my wife Jill who have tolerated
my obsession in birds for so many years,
if not uncomplainingly, almost so.

AUSTRALIAN LORIKEETS

Experiences in the field and aviary

by

STAN SINDEL

SINGIL PRESS

ISBN 0-9587727-0-3

PRINTED IN AUSTRALIA BY
SURREY BEATTY & SONS PTY LIMITED
43 Rickard Road, Chipping Norton, NSW 2170

ACKNOWLEDGEMENTS

Firstly, I would like to thank all those bird people throughout Australia whose personal observations, comments and answers to my questions have enabled me to compile and confirm the avicultural and field information contained in this book.

If it had not been for the support given and the pressure applied by Jim Gill, I am sure I would never have commenced these pages, and when I did make a start his constant comments, good advice and editing enabled me to complete them. Sincere thanks, Jim, for all your help and for your chapter on lorikeet diseases.

My thanks to Graeme Phipps for his editing, criticism and constructive suggestions for the manuscript, and for his very kind Foreword to this book.

Peter Gray's thoughts and comments on the manuscript were of particular value because they came from a non-avicultural person with an avian veterinarian background.

For the time spent editing this work by Karen Hedberg and for her sensible suggestions, particularly relating to design, I give my thanks and gratitude. Karen's help and efforts with the manuscript are of great significance to me because we have never met.

Thanks to Lance J. Ruting for his photographic advice and for his photographs.

Thanks to Ivor Beatty for his advice with the production and his firm for printing the book.

Finally, I would like to thank Marilyn Gill who gave me her favourable report on the manuscript at a time when I was full of self-doubt. Those of you who know Marilyn will understand why her approval of the work did much to restore my confidence.

TABLE OF CONTENTS

LIST OF PLATES

All photography by Stan Sindel with the exception of Plates 8D and 22C, which are by Lance J. Ruting.

AUSTRALIAN LORIKEETS

This book is the first monographic treatment of the Australian species of brush-tongued parrots in the sub-family *Loriinae.* Although members of this sub-family range throughout Australasia, Oceania and parts of Indonesia, the Australian species would appear to have the strongest captive populations. The survival of wild populations of Australian lorikeets is not threatened but other groups are not so secure. Techniques developed by people such as Stan Sindel may have profound value for the future survival of lorikeets such as some of the Pacific island forms.

In the last decade there has been a quantum leap forward in lorikeet husbandry and management (loriculture?). Stan Sindel is at the forefront of developments which have led to these advances, due to his experimental attitude to aviculture. He tries new ideas. He applies information obtained from a wide range of sources to the aviary situation. He makes major modifications to an existing set-up and routine to give new ideas "a go", and he records what happened. The ensuing breeding results from his lorikeet operation speak for themselves.

Stan has always been one who unselfishly shared ideas; in conversation, in articles written in the avicultural press over many years and in lectures — including invitations to address both national and international conventions. It is very gratifying to see him arranging his experiences and producing this book.

I both respect and admire this master aviculturist for his achievements. I share with him a fascination with lorikeets — active, noisy, colourful, yet demanding. Lorikeets are special birds with quite detailed captive requirements and this book will improve their lot by dispelling a lot of ignorance which surrounds their aviculture.

I commend the book. It is excellent.

Graeme Phipps
Curator of Birds
Taronga Zoo, Sydney
21st August, 1986

INTRODUCTION

We are living in a sophisticated, constantly changing, scientifically orientated world where every conceivable thing becomes more complicated and simplicity has become a thing of the past.

The sciences have crept into aviculture, or over the last few years, rushed would be a more appropriate description. There is no doubt that the scientific approach has made, or is in the process of making us, better aviculturists. However, do not forget what aviculture means to us. It is arguably a form of conservation, although more importantly it is a totally absorbing hobby designed to propagate aviary species and to give pleasure and satisfaction to the aviculturist. If we allow the sciences to extend into aviculture beyond the point where the welfare of the birds is uppermost in our minds, then some soul searching reform must come to aviculture. For example, scientifically organised mass production can increase the numbers of birds bred, but in the extreme case, this may be at the expense of the birds' welfare. At the same time it can deprive us of the pleasures of watching our birds rearing their young naturally.

The aim of this handbook and of the series that will hopefully follow, is to help aviculturists at all levels in the quest for further knowledge and understanding of their birds and their hobby. My intention is to keep it simple and not to complicate these pages with scientific terms beyond most of our comprehensions, yet to explain the new techniques being used in today's aviculture. Where I have used terms or words commonly met with in avicultural or ornithological literature, but which may present problems to the newcomer, they have been defined on the first occasion they are used.

The Australian Lorikeets belong to the Family *Psittacidae,* Sub-Family *Loriinae* and are members of the Genera *Trichoglossus, Psitteuteles* and *Glossopsitta.* The *Loriinae* are distinguished from other *psittacines* by anatomical developments and adaptations evolved to assist them in gathering and digesting their main food sources: pollen, nectar and fruit. Their beaks are long and narrow and the tongue has elongated papillae which form a brush-like tip adapted for the collection of pollen and nectar. Their plumage is tight and glossy and generally brilliant in colour. They are noisy, active birds with a high metabolic rate.

When the extent of the Australian continent with its huge climatic and habitat variations is considered, it is surprising that the lorikeets evolved in this region are limited to three *Trichoglossus* (the Rainbow, the Scaly-breasted and the Red-collared), one *Psitteuteles* (the Varied) and three *Glossopsitta* (the Musk, the Little and the Purple-crowned). There are only seven altogether.

Trichoglossus are medium-sized slender lorikeets with graduated tail feathers which are longer in the centre. The chest feathers are barred or scaled in appearance and the sexes are alike. The immature resemble the adults and they are as playful as young puppies.

Psitteuteles are small, slender lorikeets with graduated tail feathers in which the three members of the genus vary considerably. The only Australian member of the genus has unusual streaked plumage, a white eye ring and is sexually dimorphic (the sexes are different in appearance).

Glossopsitta is an Australian genus consisting of two small and one medium sized lorikeets with medium length wedge-shaped tails and small, slender beaks. The sexes are alike and the immatures resemble the adults.

All of the Australian lorikeets are excellent aviary subjects being beautiful, active birds with a friendly disposition. New housing and feeding techniques have simplified their aviculture. The appearance of mutations in some species is creating added interest which must ensure a bright avicultural future.

Finally, let us not lose sight of the simple uncomplicated pleasures that are derived from successful aviculture and that sense of achievement and the indescribable warm feeling we aviculturists enjoy when a healthy clutch of parent reared birds leave the nest. The ultimate aim in aviculture must be to house, feed and maintain your birds in a manner conducive to the successful reproduction of the species.

HOUSING

Due to the constant soiling of aviary walls caused by the liquid nature of their droppings, lorikeets have not always received the popularity they deserve as aviary birds. However, if some thought is given to the housing design and construction materials, they can be maintained just as efficiently as any other *psittacine*.

Designing an Aviary

When designing an aviary which is to be used exclusively for lorikeets the following points should be considered:

1. Because of the soiling effect of their droppings, solid walls and partitions should be eliminated wherever possible and replaced with wire mesh.
2. Where solid walls are unavoidable they should be of a non-porous material with an easy-to-clean surface such as colour-bonded iron sheeting.
3. Aviary frames constructed of square or round galvanised steel tubing require no maintenance and are easy to clean.
4. Lorikeets are usually quite hardy and indifferent to weather conditions, thus allowing sheltered areas to be kept to a minimum to reduce soiling.
5. Although lorikeets can be quite pugnacious with other parrots they are very tolerant of their own kind. I have found that all Australian species can be housed side by side with only single wire mesh partitions. They do not need solid or double wired mesh partitions.
6. Lorikeet aviaries, like all other aviaries, should face north in the southern hemisphere and should be sheltered in some way from the prevailing winds.
7. Some measures should be taken to prevent food and water containers from being contaminated by the birds' droppings. This is best done by having no overhead perches in the feeding areas or by having wire mesh feeding bays attached to the aviaries in such a way as to avoid fouling.

8. The minimum suitable aviary size to house a pair of lorikeets is 3 metres long, 0.9 metres wide and 1.8 metres high. Fig. 1.

9. The easiest to clean and most suitable floor material is concrete but sand or pebbles if drained with below-ground level agricultural drainage is quite effective. Aviaries with earth floors should not be used unless they are large planted aviaries with a low population density.

10. I prefer roofing at both ends of the aviary so that both front and rear perches are sheltered. This offers protection to the birds from both weather conditions and predators.

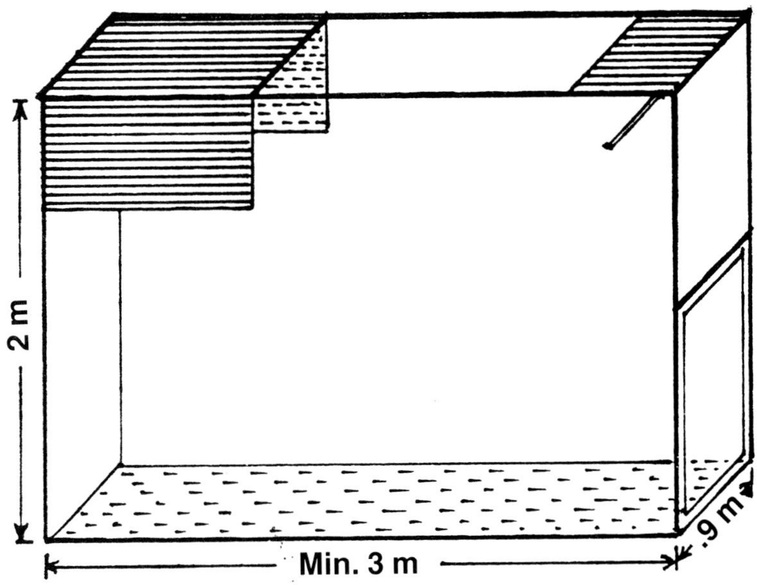

FIGURE 1: Minimum Size Lorikeet Aviary.

11. Ideally the aviaries should open on to walkways or service areas to facilitate feeding and cleaning and to eliminate escapes. When the walkway is situated on the front of the aviaries it provides further protection from predators and other external disturbances. If the walkway is attached to the rear of the aviaries and enclosed it simplifies feeding particularly in wet weather and it also facilitates the use of externally mounted nest boxes.

Large planted aviaries are suitable for housing the small species of lorikeets (Littles, Purple-crowns and Varieds), which will do little or no damage to foliage in an aviary where the plants are well established. In fact, small colonies can be kept under these conditions. I have found these varieties to be compatible with finches and doves, but some individuals are quite aggressive towards other small parrots, such as *neophemas*, when they are nesting. Pairs of Musk Lorikeets can also be housed successfully in large planted aviaries with little or no damage to plant life.

Suspended Cage Housing

In recent years suspended cage housing of *psittacines* has become quite popular, particularly in Europe and the USA. Although there may be some argument as to the advantages and disadvantages of keeping parrots under these conditions, there is little doubt that the keeping of lorikeets in suspended cages has many advantages and few if any disadvantages. They provide the birds with an environment which is hygienic, secure and allows the bird to exercise by climbing or flying. These are the essential ingredients in any aviary and any additions are usually only for human aesthetic appeal.

The cages are best situated inside a bird-proof enclosure such as an open fronted shed or inside a large aviary. Then servicing can be carried out without the fear of birds escaping and the occupants are protected from attack by predators. When the cages are situated in unsheltered areas some form of protection from inclement weather must be provided on the cage. See Fig. 2. If the cage is wholly or partially placed in a roofed and draught-free area, shelter on the cage itself may not be necessary. See Fig. 3.

The size of the cages may vary considerably. I have found that 0.6 metre wide by 0.9 metre high by 1.8 metres long is adequate, although I prefer a length of 2.5 to 3.0 metres to provide additional exercise. Batteries of cages have proved to be successful and are, of course, cheaper to construct.

The construction of cages in 2.5 cm × 1.2 cm or 2.5 cm square hot dipped galvanised weldmesh is the most suitable. The 2.5 cm square mesh has less area for soiling and although the smaller species can put their head through the mesh I have no problems arising from this.

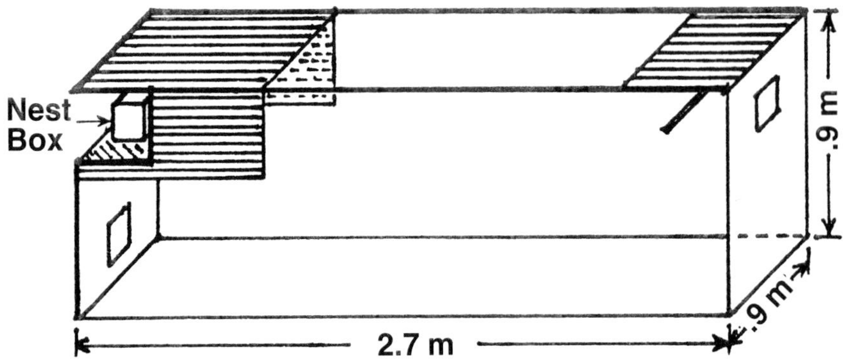

FIGURE 2: Suspended Wire Cage with Shelter.

FIGURE 3: Suspended Wire Cage for use under Shelter.

The *advantages* of suspended cages are:

1. Easier to keep clean than aviaries and when the cage is situated outside the odd shower of rain will provide almost all the cleaning that is necessary.
2. More birds can be housed in a limited space.
3. The birds kept in these cages have no access to their own or other birds' droppings and since all old food falls through the bottom of the cage as well, disease is kept to a minimum.
4. It is not necessary to worm birds housed in suspended cages.
5. Birds housed in these conditions feel more secure than when housed in aviaries, as they soon realise no one will enter their territory.
6. External nest boxes used on this cage system facilitates easy inspection and cleaning with no disturbance to the birds
7. The cost of erecting cages is far less than for aviaries.

The *few disadvantages* of suspended cages are:

1. Catching birds that are housed in this system can be a two-person job. One person with a net and another herding the birds.
2. If a suitable diet is not used some birds will become overweight which they would not do in a large planted aviary. They would also become overweight in a conventional flight aviary.
3. Efficient automatic watering systems are difficult to organise in suspended cages.

Summary

I kept lorikeets in aviaries for over 20 years before using the cage system and still house some lorikeets in aviaries. However, I firmly believe the suspended cage system of housing is the most efficient and satisfactory way to keep lorikeets. When my small lorikeets were housed on the colony system in large aviaries, there would periodically be sudden outbreaks of bacterial infections. The cage system has eliminated this. Periodic flying of cage housed birds in large aviaries must be advantageous, particularly if pairs have become overweight, lethargic, stopped breeding or are in poor feather. Many experienced and successful lorikeet breeders house their birds in aviaries and their results are such that they feel there is no need to change.

APPENDIX 1

RICE CEREAL

J. Heinz Company Australia Ltd
Princes Highway
Dandenong, Victoria E2562

Approximate Composition

100 grams of Rice Cereal contains:

Protein (N×6.25)	7.5 g
Iron	59 mg
Calcium	810 mg
Niacin	40 mg
Thiamine	2.0 mg
Riboflavin	1.8 mg
Kilojoules	1565
Carbohydrate (by difference)	78.6 mg
Fat (acid hydrolosis)	3.8 g
Phosphorus	580 mg
Crude Fibre	0.3 g
Moisture	6.8 g
Mineral Matter	3.3 g

ELLIOTT'S "EGG BISCUIT"

Chemical Analysis:

Minimum protein	15%
Crude fat	3.8%
Maximum crude fibre	1.0%
Salt	1.7%

Ingredients:

Bread crumbs
Whole egg solid

CENOVIS BREWERS YEAST

Average Analysis

5g (dried basis) of Cenovis Brewers Yeast contains:

Vitamin B1	0.5 mg
Vitamin B2	0.2 mg
Niacin	1.5 mg
Pantothenic Acid	0.3 mg
Phosphorus	120 mg
Potassium	110 mg
Sodium	5 mg
Magnesium	15 mg
Calcium	15 mg
Iron	1 mg
Copper	0.3 mg
Manganese	0.025 mg
Cobalt	25μg
Protein	2.25 g
Moisture	0.3 g
Fat	0.05 g

Vit. B6, Vit. B12, Chromium, Folic Acid, Biotin, Insotil and Enzymes have also been identified.

SUSTAGEN SPORT

Mead Johnson
345 Pacific Highway
Crows Nest, NSW 2065

	Dry Powder g/kg
Protein	245
Fat	.6
Carbohydrate	649
Ash	57
Moisture	20

Includes Vitamins A, D, E, C, B1, B2, B6, B12, Niacinamide, Pantothenic Acid, Folic Acid, Calcium, Phosphorus, Iron, Sodium, Potassium, Magnesium, per Manganese, Iodine and Zinc.

FAREX

Approximate composition

Carbohydrate	75.2%
Protein (N×6.25)	13.8%
Fat	3.0%
Minerals	4.0%
Moisture	4.0%

17 kilojoules (4 calories) per gram.
To each 100 gram has been added:

Iron	21.2 mg
Calcium	748 mg
Ohosphorus	395 mg
Thiamine	1.4 mg
Riboflavin	1.8 mg
Niacin	11.8 mg
Pyridoxine	0.8 mg
Calc. pantothenate	4.0 mg

Ingredients:
Wheat Flour, Oat Flour, Skim Milk Powder, Calcium Phosphate, Yeast, Vitamins, Iron, Salt free.

GLAXO

Glaxo Australia Pty Ltd
1061 Mountain Highway
Boronia, Victoria 3155

PENTAVITE

Nicholas Kiwi (Pacific) Pty Ltd

Each ml contains

Vitamin A	3600 I.U.
Vitamin B1	1.2 mg
Bitamin B2	1.8 mg
Vitamin B6	0.3 mg
Nicotinamide	15.8 mg
Sod. Pantothenate	0.75 mg
Vitamin C	95.0 mg
Vitamin D	900 I.U.

PANACUR 2.5

Hoechst Aust. Ltd

Fenbendazole	25 g per litre

DIETS

Early literature suggests that aviculturists of the 1870's had a good working knowledge of the dietary requirements of lorikeets, but then as now, some aviculturists insisted on feeding a total seed diet which eventually results in fits, paralysis and death. Lorikeets have received more dietary abuse at the hands of aviculturists than any other avian group. This applies particularly to the Australian members of the *Trichoglossus* genus, who because they are so hardy manage to survive for great periods of time on inadequate diets. Over the years I have been told many times, often with great pride, of some unfortunate lorikeet which has survived for an amazing period of time on a dry seed diet supplemented only with water.

Most *psittacines* can be difficult to provide with a balanced diet because of their likes and dislikes and hence their reluctance to try new foodstuffs. Lorikeets, on the other hand, will eat any form of softfood or nectar providing it is sweetened and therefore they are easy to feed and adapt to new diets. The problem with lorikeets is, of course, to know what should be included in the softfood or nectar to provide that elusive nutritious, non-fattening and balanced diet.

There are almost as many lorikeet diets in Australia as there are lorikeets; most are quite adequate but unfortunately many are not.

The requirements of a good maintenance diet are:

(a) That it maintains your birds in a healthy condition.

(b) That it does not cause your birds to become excessively fat. Trim birds breed better and live longer.

(c) That it does not sour quickly.

(d) That it does not attract insects.

(e) That it only needs to be fed once per day since many aviculturists have only a limited time to spend with their birds.

A good maintenance diet should not be adequate for parents to rear young on. If it is, the adults will become overweight when they are not breeding. It should be necessary to add a higher protein food such as egg and biscuit rearing food or madeira cake, to the maintenance diet when young are being reared.

The *Trichoglossus* Lorikeets (the Rainbow, the Red-collared and the Scaly-breasted) are hardy birds and will survive on a very poor diet, but when given a reasonable diet they will thrive.

Musk Lorikeets are not as hardy as the *Trichoglossus* genus but will also do well on a reasonable diet.

The small lorikeets (the Varied, the Little and the Purple-crowned) require a specialised diet and some dedication. These species are not suitable for beginners; some experience should be gained keeping and breeding the larger species first.

In the wild, lorikeets feed basically on pollen, not nectar as has been suggested by Gould and others. However, nectar is undoubtedly consumed during periods of nectar flow, usually in autumn or early winter. Fruits, buds, leaf tips, blossoms, insects, seed and berries are also consumed in varying amounts according to availability and the species of lorikeet. Our small species appear to be more dependent on pollen and eat no seed at all, whereas the larger species have adapted to a wide range of food types. At times insects appear to form a substantial part of a lorikeet's diet.

It should also be noted that the amino acid content of pollen is almost identical to that in whole dried egg powder. Feather defects have been induced in lorikeets by deficiencies in sulphur containing, amino acids and the defect corrected by the addition of milk powder to the diet.

I know of only one study into the dietary requirements of Australian lorikeets in captivity. This work was done by Brian Rich at Avalon, NSW, in the early 1980's. The points that I, as an aviculturist, found of interest and applied in the development of my lorikeet diet are:

1. Lorikeets appear to be able to absorb simple sugars through the wall of the crop.
2. A diet which allows the lorikeet to adjust its level of intake of nutrients by providing the food in different portions is closer to a natural diet. This is better than a composite diet, where all the ingredients are combined in the form of a sloppy porridge and the bird has no choice of the food items it requires. The split diet provides food in both a liquid (nectar) form and a dry food form (pollen).
3. The faeces of the birds fed the split diet closely resembled those of wild birds with well formed faecal stools. The birds fed the composite diet had faeces which were not well formed and resembled the diet itself and hence suggested that the birds were not digesting the food efficiently.

4. It was suggested that the protein level of 15% is adequate.

Another dietary study of Purple-crowned Lorikeets in the field was made by Churchill and Christensen in Western Australia (Australian Journal of Zoology 1970, 18, 427-37). It was observed that this lorikeet's basic diet was pollen, although during the short period when nectar flows from the blossom, large quantities of nectar were consumed. Specimens examined during the nectar flow carried a considerable amount of subcutaneous fat, particularly at the base of the throat. I have observed this same excessive fat around the crop and lower neck areas in deceased overweight aviary specimens of the *Glossopsitta* and *Psitteuteles* genera.

These few points from this comprehensive study suggest to me that captive lorikeets fed continuously on sucrose based nectar become excessively fat.

Developing a Suitable Diet

To develop a suitable diet we must first consider their nutritional requirements and, of course, this is difficult to ascertain with any degree of certainty without complex scientific testing which at present is not available to Australian loriculturists. This is why there are so many lorikeet diets and why aviculturists arrive at different conclusions.

At the risk of receiving criticism from those with more knowledge of nutrition than I, the following is the reasoning behind my lorikeet diet. Firstly, since pollen is the basic food in the wild and it is largely a dry food, then it follows that feeding a dry food in captivity is more natural. Feeding a dry food simplifies feeding as it does not deteriorate rapidly in hot or humid weather providing it is kept dry. This enables the birds to have access to fresh nutrients as they require them during the day and eliminates the need to feed the birds twice or three times a day; as recommended by some loriculturists.

My dry food diet consists of:

2 cups of Heinz rice cereal
2 cups of rice flour
2 cups of Canary egg and biscuit food
1 cup of glucose powder
1 teaspoon of multi-vitamin and mineral powder
1 dessertspoon of pollen (optional)

(The composition of these ingredients can be found in Appendix 1.)

These ingredients are mixed together and stored in airtight containers until fed to the birds. This dry food is made available to the birds at all times thus providing a constant unpolluted food source.

My next aim was to develop a nectar to provide a highly palatable and concentrated nutritional supplement in a liquid format. To avoid spoiling and obesity caused by excessive nectar, I felt that a small volume that would be consumed in an hour or so would be advantageous in the Australian climate. I collected lorikeet nectar diets from all over the world, assessed their ingredients and kept those which showed some merit, discarded the rest and added a few of my own.

The resulting nectar formula is:

1 heaped dessertspoon of	malt extract
,, ,, ,, ,,	glucose powder
,, ,, ,, ,,	Sustagen
,, ,, ,, ,,	full cream powdered milk
,, ,, ,, ,,	pollen
1 teaspoon of	brewer's yeast
½ teaspoon of	calcium carbonate powder
1 ml of	Pentavite drops or similar multi-vitamin preparation
1 can (130 gm) of	strained meat and vegetables baby food
3 cups of	water

Mix all of the ingredients in a blender or vitaminiser.

Every pair of lorikeets is fed one tablespoon (20 mls) of nectar per day.

The nectar may be stored for five days under refrigeration, or it can be frozen for longer periods of up to one month.

An additional advantage of the nectar is that it provides a ready vehicle for accurately medicating the birds.

My total daily maintenance diet for one pair of lorikeets of any species is:

1 tablespoon of nectar
½ apple
silverbeet
Add lib dry food mix

Pears, oranges, paw-paw, grapes, sweet corn and madeira cake are fed occasionally as a treat.

No seed is available to any breeding pairs of any species.

When a pair of lorikeets are rearing young both the quantity and the protein content of the diet have to be increased. I increase the nectar to four tablespoons and thicken it with one dessertspoon of Farex (see Appendix 1) and I supply as much madeira cake as the pair will eat, usually a slice about 2.5 cm thick. A dish of one dessertspoon of canary egg and biscuit rearing food moistened with one tablespoon of nectar is also appreciated. The silver beet is fed in larger quantities.

I have found that once the chicks are a few days old, the parents reduce their apple intake by at least 50%, and, in fact, some pairs stop eating apple altogether. Pairs that normally barely touch their silver beet will often consume large quantities when rearing their young. Some of my pairs of *Trichoglossus* will accept sprouted sunflower seeds and green peas when rearing young.

I have reared excellent young of all species of Australian lorikeets on this diet and some species are now into their third generation of birds being fed exclusively on this diet. It would be advantageous to have a complete nutritional analysis of this diet, but so far this has not been available. With proper analysis, small variations could be made to this diet and improve the reproductive efficiency of some of the species.

A simple diet for Trichoglossus *and Musk Lorikeets* which I have used in the past and still use occasionally for birds being held *under colony conditions*, is as follows:

A slice of brown bread is covered with one dessertspoon of Farex and two teaspoons of sugar and then made into a sop with hot water. A few drops of a multi-vitamin should be added, twice a week, after the hot mixture has cooled. This quantity of sop, together with half an apple and silver beet, is sufficient for one pair of lorikeets per day. I have reared young from the four large species of lorikeets on this diet, but I must stress it is totally unsuitable for the small species.

My last word on lorikeet diets is that if you have a diet which works well for you under your conditions — don't change it.

Hand Rearing

Every now and then due to varying reasons such as bad parents, bad luck or, as is often the case with me, due to greed, we are forced into hand rearing young. If young chicks are taken for hand rearing the parents invariably lay again and the total production of lorikeets for a year can be increased. Generally, lorikeets are good hand rearing subjects, being eager to feed and if rearing more than one, eager to keep each other clean by licking away any surplus food.

I always advise to avoid hand rearing from day one if possible, as chicks that spend their first few days with their parents have a far better start in life. If day one rearing is unavoidable, then I would suggest the following. Firstly, the newly hatched chick should be kept in some form of brooder which provides a constant temperature of 97°F and sufficient humidity so that the chick does not dehydrate. (A small container of water in the brooder will suffice.) The temperature should gradually be reduced to 90°F over the next 10 days.

The chick should be fed within two to three hours of hatching with a small amount of natural yoghurt diluted with warm water or with a small amount of fluid from the parents' crop. This step is essential to introduce the necessary bacteria into the crop to aid in digestion. From then on the chick should be fed every two hours around the clock.

My hand rearing mixture is:

1 part Heinz rice cereal
1 part canned baby apple
1 part of nectar formula described earlier

This is diluted with hot water to runny consistency and fed warm from a tiny spoon, eye dropper or syringe. After three days the around-the-clock feeds can be reduced to 5 or 6 a.m. for the first feed of the day and 10 or 11 p.m. for the last feed of the day, but still at 2 hourly intervals.

At five days of age the baby apple should be reduced by half and the rice cereal gradually replaced by a mixture of one part Farex and one part canary egg and biscuit rearing food in the case of Little, Purple-crowned and Varied Lorikeets. For the *Trichoglossus* and Musk Lorikeets, the rice cereal should be replaced by a mixture of two parts Farex, one part canary egg and biscuit rearing food and one part of 15% protein chicken crumbles. For chicks under four weeks of age, the crumbles are blended in a vitamiser. (After four weeks this is not necessary.)

As the chick's crop capacity increases and hence its emptying time increases, the feeding interval should be increased to three hourly, then four hourly and so on. They should not be given their next feed if there is food left in the crop as this predisposes to crop infections and impactions.

When chicks start to feather, heating can be discontinued in mild weather. They can be kept in a small cardboard box, covered with a towel leaving just a small air space.

I prefer to take chicks in for hand rearing at the early pin feather stage, and providing there is more than one chick and it is not the middle of winter, they do not require artificial heat (in Sydney). They are kept closely confined in a small cardboard box lined with several thicknesses of newspaper topped with a couple of thicknesses of paper towel; all of which is changed at every feed. Toilet paper is used to clean the chicks after every feed.

When lorikeet chicks are almost fully feathered they can be taught to feed themselves from a bowl. After they are eating from a bowl, they can be placed in a small cage indoors during the day and returned to their box at night. New foods such as apple, madeira cake and silver beet should be introduced to help with the weaning process.

MANAGEMENT

The most important consideration in the management of lorikeets is cleanliness. Lorikeets are susceptible to a wide range of fungal and bacterial diseases, and most of these problems can be traced to dirty conditions and stale food. Every day clean nectar and soft food dishes should be used and the previous day's dishes washed for use on the alternate day. All food and water containers must be positioned to avoid fouling from the birds' droppings.

Perches must be positioned to avoid fouling of feed and water dishes as well as other perches. The perches must also be positioned so that they allow the maximum flying distance the aviary will allow. They also need to be cleaned regularly with a good disinfectant such as the iodine based chemicals used in the poultry industry. Perches positioned under shelter need cleaning more regularly than those that are exposed to the sun and rain.

If a total moist food diet is fed it is usually necessary to feed twice a day, particularly in the warmer months, to avoid the food going sour.

All surfaces subject to soiling, such as aviary walls, floors and wire mesh, should be cleaned regularly and washed with a suitable disinfectant.

Water Containers

Lorikeets love to bathe and play, hence their water containers must be of adequate size, weight and design to prevent them from splashing out all the water when bathing or from tipping out the water in a playful romp. I have found the large plastic "D" containers which hook on to the wire mesh to be suitable for water and dry food containers when using cage housing. During the hot weather water containers should be checked twice daily.

Feeding Stations

I have found feeding stations consisting of 2.5 cm square wire mesh trays and suspended in a convenient position, to be efficient and easy to clean. They are made to whatever size is appropriate to accommodate the food dishes and the fruit. All wasted food and any droppings fall through to the aviary floor.

Floor Surfaces

Finding a suitable floor surface for lorikeet aviaries can be a problem, but I have found that concrete floors with a sprinkling of sand over them are reasonably easy to keep clean as the sand prevents the droppings and wasted food from sticking to the floor.

Pebble or lucky stone floors have a pleasing appearance and providing a good below-surface drainage system has been installed they may be cleaned easily by hosing down.

Sand filled floors which have adequate drainage to keep them dry are easy to keep clean by removing the soiled surface regularly. It is usually necessary to replace all of the sand every 12 months or so.

Earth floors require constant attention and should be avoided if possible. They need to be raked and limed regularly. Birds housed on earth floors should be wormed at least every three months.

Worming

All lorikeets not housed in wire bottomed cages should be wormed every six months. This may be carried out by catching up every bird and administering the worming agent direct to the crop using a metal crop needle and syringe. Alternatively, the drug can be mixed into the daily ration of nectar or soft food. I find this latter method to be most suitable in the case of lorikeets. The worming agent that I have found to be most efficient is Panacur 2.5, a sheep drench marketed by Hoechst in Australia. It is given at a dose of 0.5 ml for the large Australian lorikeets and 0.3 ml for the small species. It is not registered for use as a worming agent in birds and the aviculturist has to accept responsibility for using it on his or her birds.

Routine Daily Observation

It is good practice to observe all your birds at least once or preferably twice every day. Lorikeets are extremely active birds and if one should appear dull and listless there is a problem. Prompt diagnosis and treatment is essential for a satisfactory recovery.

Nesting Boxes and their Cleaning

Nest boxes are preferable to hollow logs for lorikeet breeding purposes, as they are easily cleaned and inspection is simple. A variety of nesting materials can be used such as wood shavings, peat moss, sawdust and decaying wood from rotten logs; any of which are satisfactory. I have found fine wood shavings to be readily obtainable, reasonably absorbent and easy to replace.

Cleaning of the nest box is often necessary many times during the rearing of a clutch. Some pairs will always have wet nest boxes when rearing a clutch, whereas other pairs fed exactly the same diet and provided with the same housing and nesting facilities will rear chicks with a dry nest box. Obviously the nesting material must be changed regularly in the wet boxes.

The use of larger, longer, horizontal or slightly inclined nest boxes is an advantage. In these, the babies will gather in one end of the box, usually the end furthest from the entrance hole, and use the other end of the box to deposit their body wastes. Often boxes of this type will not need cleaning during the whole rearing process. See Fig. 4.

For Scaly-breasted and Musk Lorikeets I use a nest box 22 cm (9″) square by 30 cm (12″) deep, because this size fits well into my multiple cage breeding system, and these species readily accept this sized box. See Fig. 5.

Many elaborate nest boxes with false bottoms over compartments containing absorbent materials or with gauze bottoms, etc., have been developed over the years, aimed at trying to keep the nest clean and dry. However, I have found that a simple nest box constructed of 12 mm timber, plywood or particle board with a few small holes drilled in the bottom and filled with about 7 cm of fine wood shavings gives the best results. Particle board, or pine board as it is also known, is adequate and inexpensive. The boxes can be replaced fairly economically every year or two. I have used particle board for years and have never had any problem with mortalities

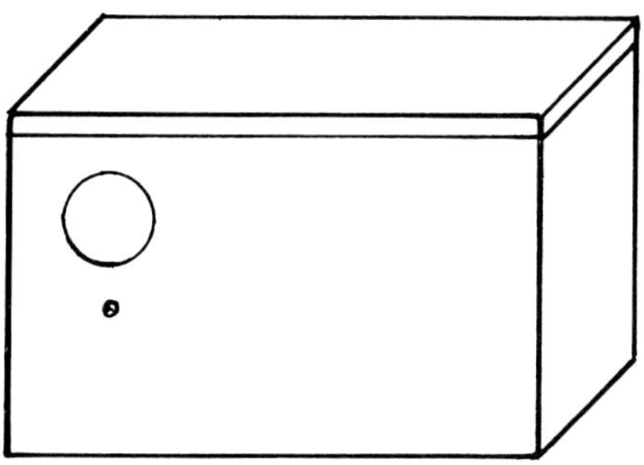

FIGURE 4: Horizontal Nest Box.

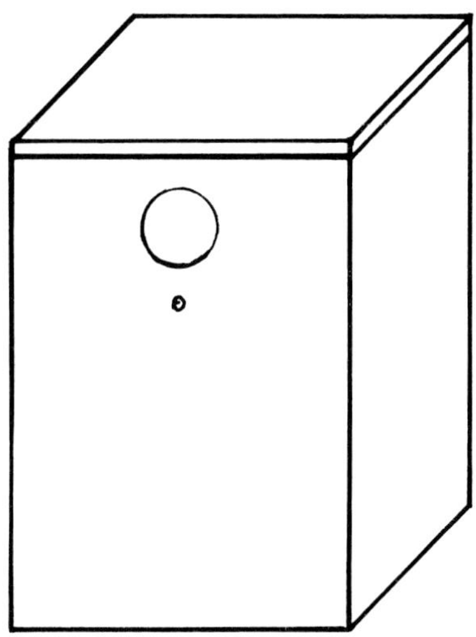

FIGURE 5: Vertical Nest Box.

associated with the production of toxic gases from the material. However, it is probably advisable to avoid using freshly manufactured particle board.

Rearing the Chicks

I have found huge variations in egg sizes between pairs and from clutch to clutch from the same pair. There is even variation between eggs within the same clutch. I believe it is a pointless exercise to record egg sizes.

When there are young in the nest, daily observation is my standard practice. If time allows, a check is made both morning and evening. Growth rate in chicks can vary greatly from pair to pair, between clutches from the same pair and between youngsters in the same clutch. However, unless something goes wrong, all the youngsters are fairly equal just prior to fledging.

Providing a suitable and adequate diet is being fed, lorikeets are usually excellent parents, but occasionally things go wrong. A chick may die and it should be removed from the nest. If a chick starts to fall behind its nest mates, supplementary feeding, fostering or hand rearing may be necessary. Hand rearing is probably the best of these alternatives as the chick may be harbouring an infectious disease and it could pose a risk to its nest mates or to its foster nest mates.

Some parent birds are messy feeders and their chicks gradually become splattered with dried-on food around the head and often extending on to the back. If this is allowed to build up, it becomes a major operation to clean off. If closed ringing is practised, the ring often becomes caked with excrement and cleaning is necessary.

All of these little problems, plus numerous others, can be overcome by constant surveillance. I have never had a lorikeet desert a nest due to my inspections, nor due to my assisting in their nesting duties.

Fostering

Should it be decided to foster rear a chick which is not doing well with its parents or because it has been orphaned, try to select a nest with chicks close to the same age. It is not necessary for the surrogate parents to be of the same species of lorikeet, as any of the four large species are capable of rearing each other's young. Likewise, any of the three small species will rear each other's young. Birds that normally only have two chicks in a clutch, such as *Trichoglossus* and Musk Lorikeets, will accept and rear at least one extra young, and quite often two.

Handling of young lorikeets can be a messy business, as they have the squirting capacity of a garden hose and what comes out is quite unpleasant. Always keep their tail directed away from you, or, if you have a similar sense of humour to myself, aim at a friend.

Egg Breaking

On a few occasions over the last five years I have had pairs of *Trichoglossus* constantly break their eggs. This problem has now been entirely overcome by providing the egg breaking pair with a large nest box 45 cm long × 22 cm wide × 30 cm deep, with the entrance hole and perch fitted to one end of the long side. A 7.5 cm internal partition is fitted on the floor across the centre of the box. The eggs are always laid at the furthest extremity from the entrance hole and the partition hinders entry to that area. It gives the brooding hen more security and provides the cock bird with somewhere to roost away from the hen (see Figure 6).

Pairing

Sexing and establishing compatible pairs in lorikeets has wasted considerable time for me in the past. Compatibility between two birds does not necessarily mean that they are a pair. I have seen very strong pair bonds develop between two males and between two females, which will not be broken even when these birds are housed with others of the opposite sex. Often the only way to break up these relationships is to separate the birds concerned and house them well away from each other. These associations can be quite confusing when trying to establish breeding pairs as the birds will feed and preen and even copulate. The dominant bird, whether it be male or female, always takes the part of the male.

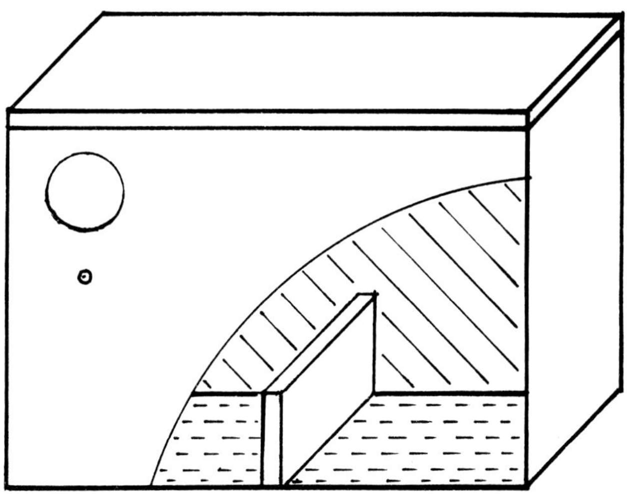

FIGURE 6: Nest Box Designed to Deter Egg Breaking.

Colony Breeding

Roger Smith, from Alice Springs in the Northern Territory, made some interesting comments on his flock breeding techniques in a lecture at the National Avicultural Convention in Brisbane in 1983. All of his lorikeets are bred on the colony system with two or three pairs per colony. He stresses that great care must be taken if new birds are introduced into an established colony, as murders can often result. To establish a new colony he takes 10 or 12 young birds and rings them all with different coloured split leg bands and places them all in one aviary. The aviaries are 4.5 m long × 2.7 m wide × 2.1 m high for Rainbows and Red-collars and 1.8 m wide for Musks and Scaly-breasted. Several nest logs are hung around the aviary at 30 to 45 degree angles from the horizontal. He uses only logs for nest sites and he avoids inspecting the nests. As soon as three pairs have gone to nest they are identified by their coloured leg bands and the rest of the birds removed. He emphasises that only established pairs should be held in a colony and that odd birds or non-breeding birds will cause disruption and possibly mutilate fledglings. Roger also pointed out that quite often the third and least dominant pair will have poor results due to the dominance of the other pairs. Feeding stations are provided for each pair as the dominant pair will often prevent the other pairs from feeding if there is only one station available.

Surgically Sexed Pairing

I have solved the sexing problem by having all of my lorikeets surgically sexed (with the exception of the Varied Lorikeet which is sexually dimorphic). Each bird is identified by placing an oversized closed metal ring on its leg whilst still under the effects of the anaesthetic used for the surgical sexing. Males are rung on the right leg and females on the left leg.

I prefer not to use split rings on any *psittacine* bird which has any reasonable strength in its beak because of the injuries they can cause themselves by closing the ring on to the leg. If the injury is not discovered quickly loss of the leg is the likely outcome.

With the problem of sexing being eliminated it is then necessary to establish compatible pairs: this can be ridiculously easy or virtually impossible. Most pairs which are placed together will settle down quickly, accept each other and eventually breed, but some will not. If surgically sexed pairs are showing no signs of compatibility after a few months they should be given new mates.

When housing breeding pairs in adjoining aviaries or cages, I have found that it is best to keep different species in the adjacent enclosure, otherwise too much interest is shown in the neighbouring pairs. Whereas breeding pairs of dissimilar species consider their aviary or cage to be their own territory and pay little attention to their neighbours.

Flock Breeding

Flock breeding is often practised in large aviaries with good results, but I believe that the ideal breeding conditions which will give the maximum results from each pair is to house them one pair per aviary or cage. When mixed species are housed together for flock breeding, hybridising often occurs. Unless there are exceptional circumstances, hybridising should be considered as undesirable, because it is important to keep species pure. Therefore, when flock breeding, it is advisable to house only birds of the same species together.

Transportation

The sensible and humane transportation of lorikeets, or any *psittacine* bird for that matter, is an area that has received little attention by animal welfare groups, wildlife authorities, bird dealers or aviculturists.

If *psittacine* birds are to be transported without constant surveillance, whether it be over long or short distances for any period of time, they should be housed one bird to a compartment. This prevents birds from perching on top of each other causing death to those on the bottom, maiming and death due to fighting and helps to eliminate stress.

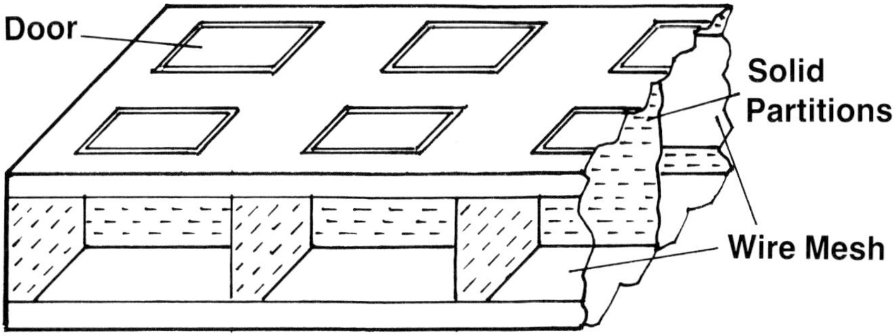

FIGURE 7: Multiple Compartment Transportation Box.

I would suggest that a suitable formula for the size of a compartment or box for a bird is the length of the bird (from beak to tip of the tail) squared by the height of the bird plus 2.5 cm. One side of the compartment should consist of at least ⅔ wire mesh (preferably 1.2 cm × 1.2 cm weld mesh). All other sides should be of a solid material with a few 6 mm air holes drilled through to assist in air circulation. If the birds will be in the boxes for no more than three or four hours I don't supply water, just some cake and a slice of apple in the case of lorikeets. Small water containers are necessary for birds being transported over long periods. Cotton wool placed in the water container will prevent the water splashing out. The foods are best placed on the floor of the compartment.

Birds should not be transported during heat wave or hot conditions unless in air conditioned vehicles.

When sending birds by air, rail or road transport unaccompanied, the box should be marked live birds, and adequate food and water provided for the duration plus a safety margin of half a day. The person they are being sent to should be notified well in advance of their pending arrival. A multiple compartment box that I have found suitable for lorikeet and small parrot transport is shown in Figure 7.

DISEASES OF LORIKEETS

James Gill, BVSc, MVM

Lorikeets as a group are subject to the same wide range of diseases as are other birds. Any bird can get one of two forms of disease — infectious or non-infectious disease. For birds to become ill with an infectious disease they must be exposed to a sufficient number of the infective agent (for example, bacteria or worm eggs) to overcome the bird's natural defence mechanisms, that is its immune system. Non-infectious diseases include such things as fractures of the legs and wings, egg binding, starvation, vitamin and mineral deficiencies. Some of the non-infectious diseases can predispose to infectious disease. The strength of the bird's immune system can be lowered by such things as cold, damp conditions, deficiencies in vitamins and minerals, lack of sufficient quantities of food to eat and constant stress from other birds, dogs, cats, mice or people including their owners.

Therefore, to minimise the chances of birds getting sick we must:

1. Reduce the birds' access to bacteria, viruses, parasites, etc. This can best be done by a month's quarantine on all birds coming into the collection and by maintaining good hygiene for all birds in the collection.

2. Make sure the birds' immune systems are working well. This is best done by ensuring that they are housed and fed well.

We should all strive to make preventative medicine the more important aim in our hobby and curative medicine a secondary role.

Lorikeets, because of the specialised nature of their diet and their liquid faeces, are particularly susceptible to bacterial and fungal diseases of the digestive system, if strict hygiene is not maintained. Because of the wide range of diseases that it is possible for them to get, we will confine this discussion to those diseases that occur commonly in lorikeets or are peculiar to lorikeets.

A large percentage of disease problems in any livestock system can be eliminated by good management. The system of housing, management and feeding as outlined by Stan Sindel in this book has gone a long way towards eliminating disease problems in those collections that have adopted it. Adult mortality has been greatly reduced and a larger pecentage of chicks are reared to maturity using these guidelines compared to housing birds in conventional aviaries and feeding them a conventional moist diet. It should be remembered that aviaries only have to achieve three things for the birds. They must provide a hygienic environment that gives the bird security and the ability to exercise. Suspended wire cages achieve these three things admirably for lorikeets. The health advantages of feeding a dry diet have been adequately discussed in the chapter on diets. The use of a small-volume liquid portion in the diet enables the lorikeets to be medicated more accurately than by using drugs in the drinking water. However, the dose given in the nectar should be calculated on the weight of the bird rather than use the dose given in water medication, so as to avoid overdosing the birds. Lorikeets have been reported to be particularly susceptible to toxicity from the antibiotics Gentamicin and Nitrofurazone (Harrison and Harrison 1986). Calcium should be removed from the nectar if Tetracyclines are used for medicating the birds. A word of warning — very few drugs are registered for use in cage and aviary birds and the situation is unlikely to change in the future. As a result all treatments and recommendations from veterinarians are based on experience at that point in time and may change as adverse reactions to drugs become known.

Feather Diseases

Feather plucking of nestlings by their parents is quite common. It is a very difficult problem to overcome. In all but the most severe cases once the chicks leave the nest they moult in normal feathers. If the opportunity arises it is advisable to select for parents lorikeets that don't pluck their young. Many budgerigar breeders believe it has a genetic predisposition. In the severe cases it is advisable to hand rear the chicks to avoid permanent damage to the feather follicles or to the wings or legs.

Psittacine Beak and Feather Disease is seen in fledgling Rainbow Lorikeets in the wild on the east coast of Australia in the late spring and summer. In those cases presented to veterinary hospitals it is characterised by loss of the primary and secondary flight feathers. The body feathers are not as severely affected as in some other species of parrots. The disease is identical microscopically to the disease in other Australian parrots and African Lovebirds. Transmission studies at Murdoch University in Western

Australia have demonstrated that the disease is associated with a virus and is therefore infectious (Australian Aviculture May 1986 pages 120-122). There is no known treatment. In contrast to the disease in other parrots where few if any birds recover, approximately 40% of wild lorikeets recover over one to two moults if they are cared for properly (Ross Perry, Personal Communication). In similar cases seen in captive lorikeets many cases also recover spontaneously.

Egg Binding

The inability of a hen to lay a fully formed egg is known as egg binding. It results in severe depression and death can be quite rapid in some cases. It has a complex set of causes but the most common factors predisposing to it are obesity and calcium deficient diets.

This is not a common problem in lorikeets which may in part be due to their overall active natures and their resultant fitness, especially of the abdominal muscles and the oviduct. Adequate calcium in the diets should make this a very rare problem in lorikeets. If a hen becomes egg bound she should be placed in a warm environment (80-85°F) and given liquid calcium every two hours until she lays the egg. Calcium Sandoz Syrup® is ideal for treatment, the larger lorikeets can be given 0.5 ml at a time, the smaller lorikeets 0.1 to 0.2 ml at a time. If this product is not available, calcium carbonate powder can be made into a suspension and the same volume given. If the bird continues to deteriorate once treatment is started or fails to lay the egg within 24 hours you should consult with your nearest avian veterinarian.

Diseases of the Digestive System

Fungal infections of the mouth and crop are very common, particularly if good hygiene is not maintained. In mild cases the only signs may be a dirty face with excess food splattered over the feathers. Diagnosis is confirmed by microscopic examination of smears from the mouth and crop. These birds respond very well to oral Mycostatin Paediatric drops®, which are available only by prescription in NSW. The more severe cases will have cheesy plaques inside the mouth and crop. In these cases the birds have often stopped eating and in the case of the small lorikeets in particular they constitute an emergency. They often require intensive supportive care from one of the increasing numbers of veterinarians with experience and expertise in cage and aviary bird medicine. This will involve careful force feeding with a crop tube, the administration of subcutaneous or intravenous fluids as well as the treatment of the causative agent with Mycostatin or a similar antifungal drug.

Protozal Infections: Intestinal coccidiosis can become a problem in lorikeets housed in conventional aviaries especially if the floors become damp. Diagnosis is confirmed by microscopic examination of the birds' droppings. They respond well to treatment with Amprolmix Plus® at the recommended poultry dose rates in the drinking water.

Worm Infections: Round worm infections are seen regularly in lorikeets housed in conventional aviaries. Diagnosis is confirmed by microscopic examination of the droppings or on post mortem examination. The treatment regime as outlined in the management chapter is adequate.

Bacterial Infections: A wide range of bacteria can infect the alimentary tract and can spread rapidly from the intestines to other organs, particularly the liver and kidneys. Bacterial infections in birds generally involve more than one system of the body. Good housing and hygiene including the avoidance of overcrowding goes a long way towards preventing the spread of bacterial disease in birds. In the early stages of the disease the birds will often respond well to the use of the appropriate antibiotics and supportive care under veterinary supervision. In the advanced stages there may already be too much damage to the tissues and organs for the bird to be able to recover despite treatment. In outbreak situations, where several birds are affected or at risk, it is often worthwhile to have bacterial cultures performed to ascertain what would be the most appropriate antibiotic to use.

A very severe form of necrotic enteritis has been seen in lorikeets in the wild and in captivity. It is characterised by sudden death and a very severe enteritis, with discrete areas of dead or necrotic tissue lining the intestine on post mortem. It is thought that it is caused by clostridial bacteria and mortalities stopped when the therapy was started with synthetic penicillins (Amoxil®).

Psittacosis is a systemic disease of birds caused by a small bacteria. Some forms of this bacteria can be highly dangerous to the birds and to people. Special stains and microscopic examination are needed to confirm the diagnosis. Treatment involves the use of long-term tetracycline antibiotics in the birds' nectar.

Viral Diseases: Lorikeets are susceptible to a wide range of systemic viral diseases. The exact nature of the diseases as they apply to lorikeets have not been studied in any detail. It should be remembered that most viral diseases are highly contagious and that there is no known specific cure. Quarantine is essential to stop the introduction of these diseases into a collection.

Surgical Sexing

This simple surgical procedure has become widely accepted in aviculture in the last few years. There are many descriptions of the operation in the veterinary and avicultural literature. It is a relatively safe, accurate and fast way of finding out the sex of your birds. Because of their high metabolic rates the small lorikeets should not be fasted prior to surgery.

In conclusion, remember that we should all try to make preventative medicine the more important aim in our hobby and curative medicine should occupy a secondary role.

PLATE 1: Size comparison showing the largest (Rainbow) and the smallest (Little) Australian Lorikeets.

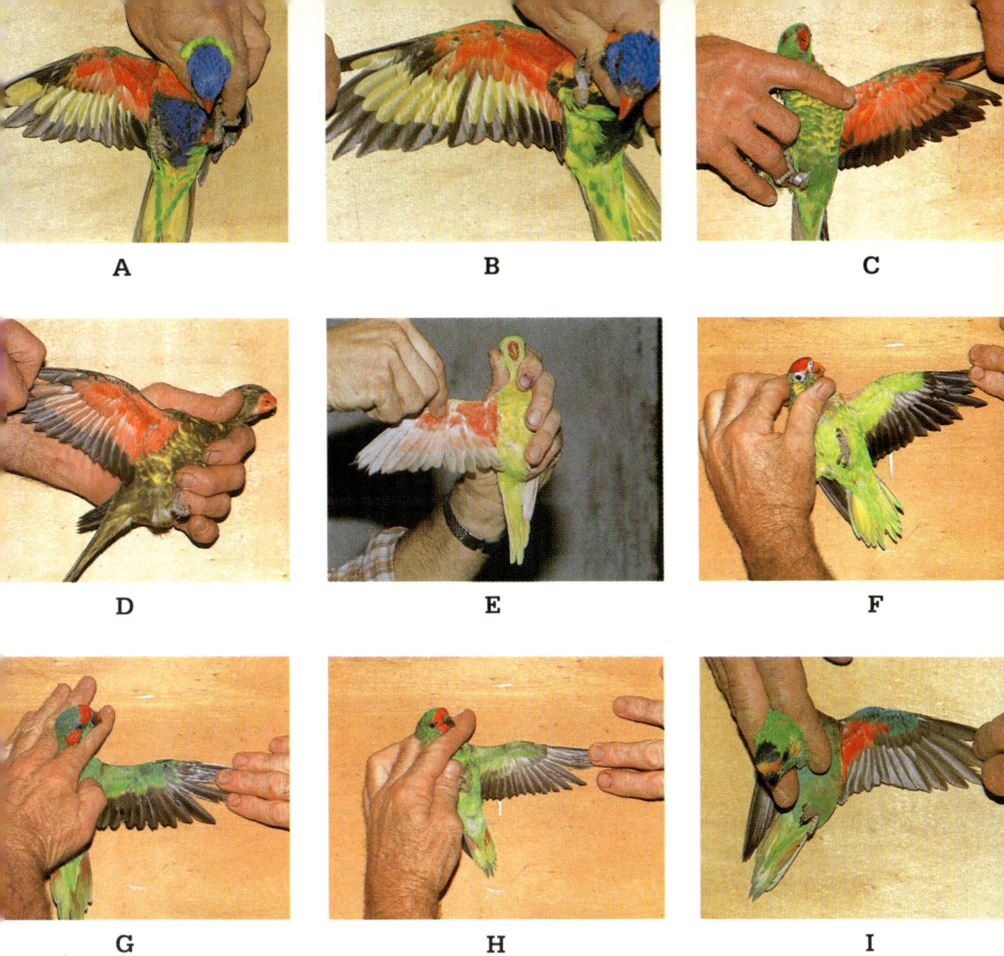

PLATE 2: Under Wing Colour Patterns.

A. Rainbow Lorikeet.
B. Red-collared Lorikeet.
C. Scaly-breasted Lorikeet.
D. Olive Scaly-breasted Lorikeet.
E. Cinnamon Scaly-breasted Lorikeet.
F. Varied Lorikeet.
G. Musk Lorikeet.
H. Little Lorikeet.
I. Purple-crowned Lorikeet.

see over page

Under wing Colour Patterns *(Plate 2):*

All Australian *Trichoglossus* Lorikeets have red feathering on the underside of the wing.

Rainbow and Red-collared Lorikeets have almost identical under wing colour patterns which appears to verify the close relationship of these lorikeets. The red is limited to the under wing-coverts and axillaries, the primary and secondary flight feathers have a broad pale yellow band, which readily identifies them when flying, and the flight tips are charcoal grey.

Scaly-breasted Lorikeets have red under wing-coverts similar to the other *Trichoglossus*, but the yellow band on the primary and secondary flights is replaced with red, thus forming the largest red under wing patch of any Australian lorikeet, which easily identifies them in flight.

Purple-crowned Lorikeets are the only other Australian species with red under wing colouring which identifies them from the other small lorikeets when in flight. The red is limited to a small patch on the under wing-coverts close to the body of the bird, and all under flights are a dark grey.

The remaining three species — Musk, Little and Varied Lorikeets — have no red and are predominantly green and charcoal grey or black under the wing. The patterns for these three species are very similar with green under wing-coverts and charcoal grey to black primary and secondary flight feathers.

RAINBOW LORIKEET

(Trichoglossus haematodus moluccanus)

PLATE 3: Rainbow Lorikeet (*Trichoglossus haematodus moluccanus*).

RAINBOW LORIKEET

(Trichoglossus haematodus moluccanus)

Derivation
Trichos — Greek for hair
glossus — Greek for tongue
haematodus — Greek for bloody
moluccanus — Latin for moluccas; originally named in error
Description
Length 31 cm
Weight 140 gm
Appearance as in Plate 3

Classification

The Rainbow Lorikeet is possibly the best known subspecies of the species *Trichoglosus haematodus* which has an extensive range from Indonesia through to New Guinea, the Solomon Islands, New Hebrides, New Caledonia, Loyalty Islands and Australia. Forshaw lists 21 subspecies of *Trichoglossus haematodus* of which two, *moluccanus* and *rubritorquis*, range into Australia. As these two subspecies are quite distinct and looked upon by aviculturists as separate species, I shall deal with them individually. Overseas aviculturists usually refer to the Rainbow Lorikeet as the Swainson's Lorikeet.

Earliest Recording

The first published illustration of an Australian parrot is of the Rainbow Lorikeet in Peter Brown's New Illustrations of Zoology, London, 3rd November, 1774.

Range and Habit

This lorikeet ranges continuously from Cape York Peninsula, down the east coast of Australia to Tasmania and along the southern coast to the Eyre Peninsula. An isolated population exists in the Perth district of Western Australia, which probably originated from aviary escapees. The range of the Rainbow Lorikeet is by far the most extensive of any of the Australian lorikeets and extends further south than any other *Trichoglossus.*

There is considerable colour variation, particularly in the red, yellow and orange colours of the chest. These variations have no bearing on age, sex, diet or the locality they inhabit. In the past many subspecies were recorded and described, but all of these are now considered to have been colour variants.

Birds from the northern part of the range are smaller and have a brighter, more evenly coloured blue head and abdomen, and until recently were recognised as the subspecies *Trichoglossus haematodus septentrionalis*.

Naturally, a species with such an extensive north-south range has a vast variation in habitat, from rainforest to the dry, open eucalypt forests, mallee scrub and mangrove swamps; in fact, anywhere there is a suitable food source. Their diet consists of pollen, nectar, fruit, seed, vegetable matter and insects. They have proven to be very adaptable with regards to introduced food sources, at times causing considerable damage to ripening grain crops and orchards. This adaptability to new foods has made them an extremely successful species, quite capable of co-existing with man.

The Rainbow Lorikeet, like all Australian lorikeets, is mainly nomadic although resident populations have been established in good food source areas, particularly around our east coast towns and cities where natural foods are supplemented by the blossom, fruit and seed of introduced flora. It has been a common resident species in the northern suburbs of Sydney for many years and more recently has extended its range into the southern and south-western suburbs. I often see small flocks in the south-western suburbs where five years ago they were unknown.

Records show that Rainbow Lorikeets had become uncommon around Sydney by the 1930's, where they were formally quite numerous. Fortunately, their numbers have increased since then.

I have observed this species throughout its entire range and have been unable to detect any variation in their behavioural patterns. With the exception of the birds in the far north, I feel there must be considerable integration throughout the range.

Breeding in the Wild

Breeding usually takes place in the spring and summer but has been recorded at most other times of the year, due no doubt to the huge variation in the climatic conditions throughout its range. The nesting site is usually a hollow limb or hole in the tree, usually a eucalypt, at heights around 16 metres (50 feet). The hollow is prepared by both the male and the female. The usual depth is 45 cm to 60 cm (1′ 6″ to 2′). They lay two and occasionally three white eggs, usually on a base of chewed up rotten wood and incubation is carried out by the female only.

An unusual nesting in the Gosford district of NSW has recently been brought to my attention, in which a pair is nesting in a hole in the trunk of a red gum tree only three metres above the ground. The tree is situated in swampy ground, the nest entrance is via a knot hole and the nesting chamber is 22 cm in diameter and 45 cm deep. Two eggs were laid in this hollow in June 1986 and were hatched in early July. Nesting continued successfully until the record rains of early August when the chicks were drowned in the nest. The observer, Robert Adams of Empire Bay, NSW, who supplied me with this information, has since found another pair nesting in a hollow five metres above the ground. This data only confirms that there are exceptions to every rule in bird behavioural patterns, both in the wild as well as in our aviaries. Winter breeding is quite unusual in this locality, as are the low nesting sites.

Aviculture

Rainbow Lorikeets have been known in captivity since 1771 when Joseph Banks took a live specimen to England and others have been recorded in captivity as early as 1783 (Dr Karl Russ, The Speaking Parrot, 1884, Page 180). The first recorded breeding has been credited to Mr A. Heublein of Coburg, Germany, in 1873. This breeding was followed by many other European successes within the next few years. Although there must have been many early Australian breeding successes, they appear to have gone unrecorded. Alan Lendon stated in the Aviculture Magazine in 1944 that the first breeding in Australia was by Dr W. Hamilton of South Australia, who was awarded the Avicultural Society of South Australia Medal for his success in 1929.

Diet

When this species is fed on a good balanced diet and housed adequately, it can be a prolific breeder. There are many records of four clutches of two chicks per clutch being reared in a 12-month period.

Sexing

Visual sexing is difficult and by no means reliable. males usually have a larger, broader head, but not always. Colour or behaviour is no indication of sex; often two males or two females will give the impression that they are a pair. The only sure, quick way of obtaining a true pair is to have them surgically sexed and this can save many wasted seasons.

Display

The courtship display of the cock Rainbow is quite spectacular. He approaches the hen with a bobbing and bowing motion and stretches to his full height with his neck arched. The pupil of the eye is constantly dilating and constricting and he emits a low, single-note whistle. The amount of interest shown by the hen depends on how near she is to nesting.

△ **PLATE 4:** Yellow Rainbow Lorikeet (possibly an acquired colour).

◁ **PLATE 5:** Fifty percent Olive Rainbow Lorikeet Hybrid.

Nest boxes

Rainbow Lorikeets are not hard to please as far as nesting sites are concerned. They will accept hollow logs, either horizontal or vertical, and nest boxes of various types. They have often been known to nest on the ground in a corner or behind some object which affords some privacy. Some have even dug a burrow when suitable nest sites were not available.

I have found nest boxes are the most suitable, as they are easy to clean and maintain. A larger box is preferred and I favour a box approximately 45 cm long, 22 cm wide and 25 cm deep, with a 7.5 cm wide entrance hole and a perch at the top of the long side. See Fig. 4.

Nesting and Hatching

Two eggs form the usual clutch in aviaries, as in the wild, but three eggs have been recorded. The interval of time between the laying of the eggs varies considerably from pair to pair and from clutch to clutch. I have recorded intervals of two, three, four and eight days between eggs in nests where both eggs have hatched. Incubation is carried out by the female and usually commences with the first egg, but some hens will not commence incubation until one or two days after the first egg has been laid, or in odd cases not until the second egg is laid. Some cocks will spend a little time in the nest during the day while incubation is in progress and most cocks sleep in the nest at night.

I have found the incubation period is variable possibly due to the weather conditions and how well the individual hen sits during the early stages of incubation. The incubation period should always be calculated on the second egg, as it is reasonable to assume that incubation would start immediately the second egg is laid. My records show incubation periods range between 22 and 25 days with the most usual period being 23 days.

Chicks hatch with a wispy white or silver down. At 10 days of age a grey secondary down starts to appear, the eyes open at 14 days and at 18 days the chicks are covered with grey down. Pin feathers start to appear about 22 days, usually on the head first. The tail and the flight feathers start to show at 28 days and the chest and abdomen feathers start to show a couple of days later. Chicks are almost fully feathered at 45 days and I have records of chicks fledging from 57 to 64 days with 60 days being the most usual.

When the young leave the nest they are slightly smaller, duller replicas of their parents with a dark brown upper and lower mandible, with some chicks showing an orange wash through the brown. The cere and the skin area around the eyes is a pale grey.

Both parents continue to feed the young for two to three weeks after leaving the nest, but they can be seen sampling the food about one week

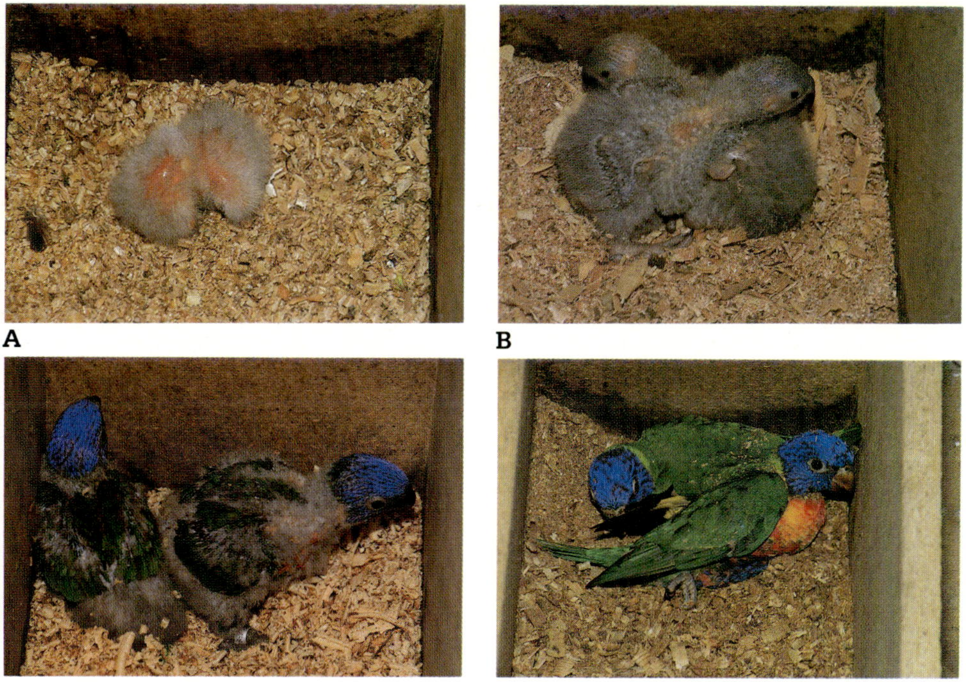

A B C D

PLATE 6: Rainbow Lorikeet Nestlings at 1 day, 3 weeks, 5 weeks and 7 weeks old respectively.

after flying. I find the young can be taken from the parents three weeks after fledging. The moment the parents show any aggression towards the young they should be removed, even if supplementary feeding is necessary.

The beak colouring has usually changed to a deep orange by four months of age and the eye colour has become similar to that of the adults.

Although Rainbows appear to mature at eight to nine months of age, I have had little success breeding from birds under 18 months or two years old.

Mutations

In Australia, an unusual golden yellow mutation form has appeared, which does not seem to conform with the laws of genetic inheritance. At least two of these golden yellow birds are known to have moulted in the yellow over a number of years. Until more work is carried out on these birds, their true potential cannot be gauged. See Plate 4.

I have received information from the USA that a cinnamon mutation has been developed in South Africa.

By hybridising the Rainbow with an olive Scaley-breasted Lorikeet, I have produced olive hybrids. See Plate 5. If I can breed these hybrids back to Rainbows and then the resulting olive hybrids back to Rainbows again, Olive Rainbows should be produced. This mutation is dominant.

Hybrids have been recorded with Red-collared, Scaly-breasted and Musk Lorikeets. All hybrids with Australian lorikeets have proven to be fertile.

RED-COLLARED LORIKEET

(Trichoglossus haematodus rubritorquis)

PLATE 7: Red-collared Lorikeet (*Trichoglossus haematodus rubritorquis*).

RED-COLLARED LORIKEET

(Trichoglossus haematodus rubritorquis)

Derivation

Tricho — Greek for hair
glossa — Greek for tongue
haematodes — Greek for bloody
ruber — Latin for red
torques — Latin for collar

Description

Length 30 cm (12 inches)
Weight 120 gm average
Appearance, as in Plate 7

Classification

The Red-collared Lorikeet is another subspecies of the numerous and highly successful species *Trichoglossus haematodus.* Although closely allied to the Rainbow Lorikeet, its isolation from this subspecies and other members of the genus has allowed it to evolve with distinct colour patterns.

Earliest Report

Early records of this species is limited. It was described by Vigors and Horsfield in 1827 and was depicted in Edward Lears' "Illustrations of the Family of *Psittacidae*" in 1832. John Gould quotes from Gilbert's field notes in "Birds of Australia" in 1842.

Range and Habitat

The range extends across northern Australia from the vicinity of Broome, Western Australia, northward and eastward to a point close to the Northern Territory and the Queensland border in the Gulf of Carpentaria. This lorikeet has been kept distinct from the Rainbow Lorikeet by the dry, treeless plains which extend from the southern base of the Gulf of Carpentaria, inland for some 200 kilometres.

There appears to be little variation in this bird throughout its range, which could be termed an east-west range. This is in comparison to the Rainbow Lorikeet which has an extensive north-south range and varies in size and colouration.

This lorikeet is numerous throughout its range wherever a suitable food source is available. I found it to be the most commonly seen *psittacine* in the

Northern Territory, usually found in the open forest where eucalypts are flowering. On a recent trip to the Northern Territory (July 1986) which included the country from Darwin to Katherine, through Kakadu National Park, Arnhem Land and up to the Coburg Peninsula; I found this lorikeet to be numerous everywhere. It was usually seen in pairs or small flocks in habitats, which included dry open forest, tropical forest, monsoon forest and paperbark swamps.

Its feeding habits are similar to the Rainbow Lorikeet and field sightings and stomach content analysis shows that it feeds on pollen, nectar, vegetable matter, fruit, seed and insects. A liking for domestic tropical fruits has developed, particularly mangoes. I was told, while in the Territory, of flocks of Red-collared Lorikeets which visited a cattle station annually when the mangoes were ripening. They would gorge themselves for weeks and as the last of the fruit became over-ripe and fermented in the hot sun, the lorikeets would become so intoxicated that they would be unable to fly and would wander around under the trees in a stupor. During July, 1986, in the Kakadu, Arnhem Land, and the Coburg Peninsula area I found them feeding mainly on the beautiful bright orange blossom of *Eucalyptus miniata*.

This species is nomadic and its movements are totally related to the food supply. I can find no definite evidence of resident populations having established as with other members of the *Trichoglossus*, but from information I have received I feel there is a possibility of some permanent residents around Darwin.

Breeding in the Wild

Hollow limbs and holes in trees, particularly eucalypts, are the usual nesting sites at varying heights from 5 m to 30 m. The nest is prepared by both the male and the female. The clutch of two eggs is incubated by the female and the male roosts in the hollow with the hen at night. They are normally spring and summer breeders, but autumn nests have been recorded. Availability of food and rainfall must have some effect on the breeding activity of this northern species.

Courtship display is similar to that of the Rainbow Lorikeet.

There is no colour difference between the sexes. Males often have a larger head and beak which can be used as a guide, but not as a definite indication of sex.

Aviculture

Avicultural records indicate that Red-collared Lorikeets were first exported to England and Europe late last century and the first reported breeding appears to have been in London's Regent's Park Zoo in 1910. The first recorded breeding success in Australia appears to be that of Mr S. Harvey of Adelaide, South Australia, in 1934, for which he was awarded the South Australian Avicultural Medal. There is little doubt there are many unrecorded breeding successes with this species prior to this date.

Diet

The avicultural requirements of this lorikeet are the same as for the Rainbow Lorikeet. Since becoming established as an aviary species they have proven to be just as hardy and prolific as the Rainbow.

Colony Breeding

Some years ago I had a colony of eight or ten birds, which were not even pairs, housed in an aviary about 15 metres (50′) long, 3 metres (10′) wide and 2.4 metres (8′) high, together with several parakeets and some large softbills. They bred constantly under these conditions and tolerated the other occupants of the aviary. Looking back I wonder whatever induced me to break up and dispose of this colony. Perhaps it was the constant soiling of the aviary walls.

In passing I would like to offer some advice for successful breeding of colonies of birds, no matter what the species. Over the years I have had several such colonies and have seen numerous others. All have worked well until something upsets this balance. Never add to or remove from a successful colony, except the young. Never try to relocate the colony or refurnish the aviary and pray you never lose a key bird in the group. The whole success usually depends on everything remaining the same.

Nest Boxes

Large nest boxes are preferable, providing the birds will accept them. See Fig. 4, as suggested for Rainbow Lorikeets.

Display

Courtship display is similar to that of the Rainbow Lorikeet.

Nesting and Hatching

The clutch size is usually two eggs laid at a two- to three-day interval. I can find no records of this bird laying more than two eggs. My records show an incubation period varying from 22 to 25 days with the usual period being 23 days. The fledging time is variable but averages 61 days.

Like the Rainbow Lorikeet, the chicks hatch with a silvery white down and their development in the nest is similar to this species.

Young birds have a dark brown to black beak with some showing an orange suffusion when they leave the nest. The eye colour is a dark brown and the plumage is a slightly duller replica of the adults, but usually with a

PLATE 8: Red-collared Lorikeet Nestlings at 1 day, 7 days, 5 weeks and 9 weeks old after fledging. Note dark beaks of fledged birds.

less clearly defined and narrower collar. By three months of age the beak has changed to reddish orange and the eye colour is almost the same as the adults.

At 12 months of age they appear to be fully mature although breeding successes under 18 months to two years of age are rare.

Mutations

I can find no record of any mutation in this species.

Hybrids have been recorded with the Rainbow, Scaly-breasted and Varied Lorikeet; as well as several exotic lorikeets.

SCALY-BREASTED LORIKEET

(Trichoglossus chlorolepidotus)

PLATE 9: Scaly-breasted Lorikeet (*Trichoglossus chlorolepidotus*).

SCALY-BREASTED LORIKEET

(Trichoglossus chlorolepidotus)

Derivation
Tricho — Greek for hair
glossa — Greek for tongue
chloros — Greek for yellow-green
lepidotus — Greek for scaly

Description
Length 24 cm (9.5 inches)
Weight 85 grams average
Appearance as in Plate 9

Classification

This species is a typical member of the *Trichoglossus* genus which shows no variation throughout its range and with no subspecies being recorded.

Earliest Report

This species was neglected by our early naturalists and was not described until 1820 by Kuhl and was first portrayed by Stephens in Shaw's "General Zoology". Edward Lear's illustration in "Illustrations of the Family *Psittacidae*", 1832; and listed under the correct scientific name and common name of Maton's Parrakeet leaves a little to be desired.

Range and Habitat

The range extends over most of eastern Australia from Cooktown in the north to approximately 100 km south of Sydney, and extends further west than other eastern lorikeet, along the watercourses of southern Queensland and northern New South Wales. They are more numerous in the centre of the range and gradually decrease in numbers towards the northern and southern extremities. There is no variation throughout the range, except that birds from the northern extremes are smaller.

An isolated population has developed around Melbourne, possibly from aviary escapees. Recently I watched Scaly-breasted Lorikeets feeding from a food tray in the company of Rainbow Lorikeets in a Melbourne suburb.

This lorikeet is mainly nomadic, but like the Rainbow Lorikeet, it has established resident populations, particularly in Brisbane and the Gold Coast areas. Fellow aviculturist and friend Alan Lynch, who resides on a small acreage 12 km from Brisbane GPO, has fed lorikeets on his property for many years. He informs me that the flocks which gather to be fed every afternoon are almost always predominantly Scaly-breasted Lorikeets together with lesser numbers of Rainbow Lorikeets. The proportions may vary from 80% Scalys down to occasionally 50% Scalys. The numbers of Scalys fed, range from in excess of 1000 down to 100 birds. He believes that the 100 birds are permanent residents in the area and the large fluctuations are due to arrivals and departures of nomadic flocks.

Breeding in the Wild

Usually a spring and summer breeder, there have been some records of nestings in the winter. In many localities breeding coincides with the flowering of the eucalypts. Nest sites are hollow limbs or holes in the trees, often 20 metres (66′) from the ground although nests as low as three metres (10′) have been recorded. This species seems to prefer its nesting chambers in tree limbs, sometimes two metres (6′ 6″) from the entrance hole. The chamber is prepared by both the male and the female. Two white eggs are laid on a base of chewed-up decaying wood and incubated by the female.

The Scaly's feeding habits are similar to the Rainbow Lorikeet, and both species are often seen travelling and feeding together. The Scaly-breasted Lorikeet will often attack ripening grain crops when the seed heads are at the milky stage.

Aviculture

There were importations of this lorikeet into England and Europe in the early 1880's. Seth-Smith reports a large consignment received by an English dealer in 1883. The first breeding was by "H.J." recorded in "Notes on Cage Birds" in 1899. This record probably refers to the successful unconfirmed breeding by a Mr Johnson, headmaster from London, in 1884. The next recorded breeding was by Berlin Zoo in 1890 where the pair were housed in a one-metre cubed glass cage. They laid two eggs in a nest box and two young were fledged at two months of age.

In Australia the early avicultural successes with this species, as indeed with most other species, have gone unrecorded. The first recorded breeding appears to be that of W. K. Penny in 1935, for which he was awarded the Avicultural Society of South Australia Medal.

The aviculture of this species is similar to the other *Trichoglossus*. They will breed reasonably well in a flock situation, but the best results are obtained when single pairs are housed in aviaries or suspended cages.

Sexing

Male and female have similar colouring, but males often have a larger, bluer head which gives a guide, but is by no means a definite indication of sex.

Display

I have found the Scaly-breasteds' courtship display to be a little different from the other Australian *Trichoglossus*. The male approaches the female stretched to his full height with neck arched and pupils dilated. He hops along the branch and often turns in a full circle. While all this is happening, he utters what is best described as a musical "coo".

Nest Boxes

For convenience I use a nest box 22 cm (9″) square by 30 cm (12″) deep, because this size fits well into my cage battery breeding system and this species readily accepts this size of box. The larger the horizontal area of the nest box, the less cleaning required when the young are being reared, but some pairs will resist very large nest boxes. See Fig. 5.

Nesting and Hatching

In captivity, two eggs, very rarely three, are laid at a two- to three-day interval, and incubation may start any time between the laying of the first and second egg.

My records show incubation periods of 22 days and 23 days with the vast majority of eggs hatching at 22 days. The fledging period in my aviaries has been from 56 to 60 days, with the most common period being 57 days.

Chicks hatch with silvery-white down and the grey secondary down starts to appear at 10 days of age. Eyes are open by 14 days and pin feather development is visible about 20 days.

Young birds leave the nest with dark brown to black beaks and in some the beaks are suffused with orange. The eye is dark brown. The plumage is similar to, but a slightly duller colour than that of the adults and they show less yellow behind the neck and on the upper mantle. Within three months of hatching the mandible has changed to orange-red in colour and the eye colour is similar to that of the adult.

When they are 12 months of age they are identical to adults, although breeding success is unusual before they are 18 months old.

△ **PLATE 11:** Cinnamon Scaly-breasted Lorikeet. Most probably sex-linked mutation.

◁ **PLATE 10:** Olive Scaly-breasted Lorikeet (a dominant mutation).

Mutations

I can find records of more colour mutations occurring in this species than any other Australian lorikeet. The first recorded mutation to occur was a blue. A striking blue and white scaled bird with a reddish beak. At least two of these birds have occurred: one is a museum skin and the other was held in Sir Edward Hallstrom's collection over 20 years ago. This bird passed into the hands of another aviculturist, but unfortunately the mutation was never established. Rosemary Low refers to a blue mutation in her book "Lories and Lorikeets", but whether it is the Hallstrom bird, or yet another bird, I do not know.

The olive mutation appeared first in Brisbane about 15 years ago and after a couple of shaky starts it now appears to be establishing itself as an aviary strain. I have found it to be a totally dominant inheritance, which has no intermediate colour (as is the jade in the olive Peachface African Lovebird). There are no visible single and double factors, as in some dominant pied mutations. See Plate 10.

A jade or laurel green mutation has recently occurred in Queensland, which has no relationship or bearing on the known olive mutation. As yet very little work has been carried out on this mutation, so it is too early to ascertain if it is a single factor form of another distinct olive mutation, or it is an intermediate colour mutation between the normal and olive mutations.

A cinnamon mutation bred from two normal birds appeared in a northern NSW aviary about three years ago. Since then two more have appeared, but died before fledging. I recently examined one of these birds which had been frozen. It was a beautiful three-quarter feathered chick that was being hand reared when it suddenly died. This mutation hatches with red eyes and appears to be a typical sex linked cinnamon, although the genetic inheritance is unconfirmed as yet. See Plate 11.

Another cinnamon mutation was bred by John Fraser of Bowen, Queensland, a few years go. The parents were normal greens, except the female had a yellow beak. Later, the cinnamon, its parents and normal birds from the same pair, were purchased by another Queensland aviculturist, but unfortunately things did not go well. The cinnamon and the yellow-beaked hen have since died. The original green cock has been paired with a new hen, but as yet only green youngsters have been produced. I am told that this cinnamon appeared to be identical to the previously mentioned cinnamon mutation.

In my experience, when normal *Trichoglossus* lorikeets lighten in the beak colour, either to a pale orange or yellow, it is usually associated with poor health. Perhaps the yellow-beaked hen was, in fact, sick.

It is quite common to find Scaly-breasted Lorikeets in aviaries and in the wild with varying amounts of yellow feathering, often showing 75% yellow. This colour variation is usually related to a blood disorder or a vitamin deficiency and invariably moults out when the bird is fed a good balanced diet. As yet I have not seen a true pied mutation in any Australian lorikeet, but hopefully one will eventually occur.

PLATE 12: Immature normal and olive Scaly-breasted Lorikeets, the result of pairing an olive male to a normal female.

A

B

C

D

PLATE 13: Scaly-breasted Lorikeet Nestlings at 2 days, 10 days, 4 weeks and 8 weeks old respectively.

VARIED LORIKEET

(Psitteuteles versicolor)

PLATE 14: Varied Lorikeets (*Psitteuteles versicolor*). Note the brighter and more extensive chest colour of the male.

VARIED LORIKEET

(Psitteuteles versicolor)

Derivation

Psitte — Greek for parrot
euteles — Greek for paltry
versicolor — Latin for different coloured.

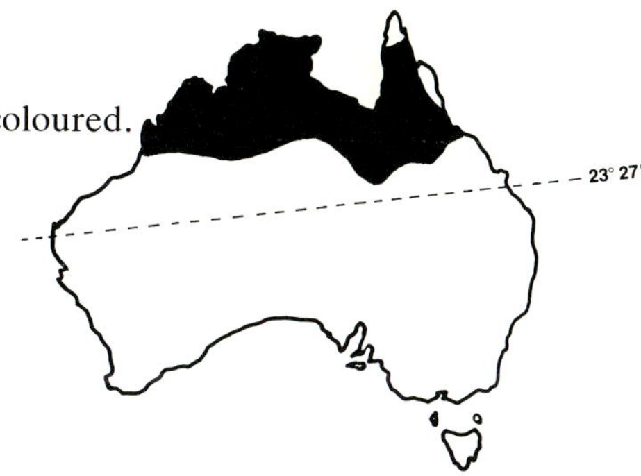

Description

Length 19 cm
Weight 60 gms
Appearance as in Plate 14.

Classification

Recent authors have reclassified this species. Joe Forshaw includes it with the *Trichoglossus*; Rosemary Low classed it as a *Glossopsitta.* I believe that the Varied Lorikeet has no close affinities with any other genus and should be installed in a monotypic genus. Aviculturists have a close and personal relationship with the birds in their care and from this intimate contact we can often offer more information on the relationship of one species to another than can the taxonomist. Both Rosemary Low and Joe Forshaw are themselves aviculturists, but neither have ever kept Varied Lorikeet in their collections.

The following are the features that I feel set this species apart from the *Trichoglossus* and the *Glossopsitta.* Firstly, it is the only lorikeet with a marked white periopthalmic eye ring, its feather patterns are unlike that of any other species and it is sexually dimorphic unlike the *Trichoglossus* or *Glossopsitta. The Varied lays three to five eggs unlike the Trichoglossus* and the Musk Lorikeet; its young fledge on an average of one week sooner than *Glosssopsitta*, and two weeks sooner than *Trichoglossus* and its chicks hatch with less down than any other Australian Lorikeet. In addition, their behaviour is different to the other Australian lorikeets in aviaries; they do not become as friendly as the other lorikeets, they remain aloof and are somewhat sluggish and prone to becoming overweight.

I have left the Varied Lorikeet in its original genus, *Psitteuteles*, along with the Goldies and the Iris Lorikeet, not because I feel that is where they should be, but because I don't feel qualified to change this.

Earliest Records

This lorikeet was virtually unknown until Lear illustrated it in 1831. Gould was the first to publish field observations, in his "Birds of Australia", Volume 5, 1842; from field observations supplied by John Gilbert of his sightings at Port Essington.

Range and Habitat

The range of the Varied Lorikeet extends across northern Australia from the Kimberley area of Western Australia to the east coast of northern Queensland. It is interesting to note that the dry, treeless plains which extend southward from the Gulf of Carpentaria, and which have effectively acted as a barrier to divide the Red-collared and the Rainbow Lorikeet, have not effectively divided the range of the Varied Lorikeet. It would be interesting to know how much movement, if any, there is between the eastern and the western populations.

I have had the opportunity to observe this species in Cape York, the Gulf country and the Northern Territory, all during the winter months. My impression is that the Cape York population is far more scattered and congregates in smaller flocks (usually 10 to 20 birds) than the birds from the west of the range. Perhaps this suggests that the Cape York population has little or no contact with birds from the rest of the range. Cayley states that the birds from north Western Australia are deeper and brighter in colour than the birds from Cape York. Mathews recorded two subspecies, but I feel these variations require more investigation and probably only relate to individual birds.

The Varied Lorikeet could be termed migratory in the Gulf Country or more precisely in the Argylla Ranges which lie between Cloncurry and Mt Isa. Their arrival in this district, usually in June, coincides with the flowering of the bloodwood trees (*Eucalyptus terminalis* and *Eucalyptus polycarpa*) followed by the Bauhina and the Melaleucas. An old friend, Don Payne, from Cloncurry, assures me this has happened every year for as long as he can remember, a period spanning 50 years. The birds then breed, laying their first clutch of eggs in early July; the youngsters fledge in late August or early September. In a good season when the blossom is plentiful a second brood is often reared. I have seen flocks of thousands of these lorikeets, made up of 75% immatures, screeching and zooming through the river gums along the dry watercourses. When in flight, this species of bird gives the impression of being much larger than it actually is.

In 1980, when the bloodwood trees flowered a month earlier than usual, the lorikeets arrived and bred a month earlier than usual. The feeling of the local residents in this area is that the Varieds (or Red Caps as they are known locally) migrate from the Northern Territory, but this is by no means proven. How these birds co-ordinated their arrival with the early flowering of the bloodwoods is not understood.

Basically, this species is nomadic although there does appear to be a resident population in the Darwin area, similar to the resident populations of *Trichoglossus* around the eastern capital cities. However, certain sections of the population do apparently migrate to the Argylla Ranges annually.

John McKean, ornithologist and renowned bird observer of Darwin, informs me that he has recorded Varied Lorikeets in every month of the year in the Darwin area but has never recorded them breeding in the area.

In the Argylla Ranges I have usually found this species along the tree-lined dry watercourses which are common in the area. They disperse a kilometre or so from these watercourses, across the arid terrain to the odd isolated flowering tree. On Cape York I saw them in small parties in dry, open, tree-studded country bordering flowing watercourses. In the Northern territory, I found them in dry forest country, in large flocks, often in the company of Red-collared Lorikeets.

While on a recent bird watching trip to the Northern Territory (July 1986) which extended from Darwin to Katherine then through Kakadu National Park, Arnhem Land and to the Coburg Peninsula, I found this lorikeet to be numerous in all localities we visited. They were constantly seen in small flocks of up to a dozen or so birds and in Kakadu and on the Coburg Peninsula they were often seen in flocks of 100 or more birds. No immature birds were observed in any of the flocks which indicated that breeding had not yet commenced. I felt that it was not far off as there were many pairs within the flocks sitting side by side engrossed in mutual preening.

The Varieds were feeding on the blossoms of various eucalypts, particularly the bloodwood and the beautiful deep orange blossom of *Eucalyptus miniata*, as well as Melaleucas. I also watched them gleaning insects from the eucalypt foliage.

Habitats noticed to be frequented by this lorikeet included dry, open forest, tropical forest, monsoon forest and Melaleuca swamps.

The forests in the Coburg Peninsula provide very few nesting hollows suitable for the large numbers of this lorikeet seen in the area to breed. This indicated to me that the majority of these birds would move inland to the dryer open forest country which affords more nesting hollows to breed in.

A B C D

PLATE 15: Varied Lorikeet Nestlings at 1 day, 1 week, 3 weeks and 4 weeks old respectively.

Breeding in the Wild

Nest sites usually are in hollows in eucalyptus trees. The entrance is usually a small knothole or spout. Most nests I have observed have been in limbs angled from horizontal to 45 degrees. These nests, which are prepared by both the male and female, are often in excess of 100 cm deep. Three to five eggs are laid in decaying wood or on a termite nest which are common in live trees in parts of northern Australia. I once found a large spreading eucalypt, standing alone, some 150 metres from a dry creek, which had three Varied Lorikeet nests and one Cloncurry Parrot nest within its branches. All were feeding young and making regular excursions to and from the tree. The lorikeet nests were well apart from each other and each pair ignored the presence of the others. I witnessed no fighting between the pairs.

Probably due to the climatic instability and the wide variation in habitats in northern Australia, nesting has been recorded in all months of the year. However, the usual and most favoured breeding time is the northern dry season, from April to September.

I have observed young birds in the Argylla Ranges many times and they leave the nest similar in colour to the adult female, but without the red crown and with a brick-red suffusion on the face. In Forshaw and Cooper's second edition of "Australian Parrots", the illustration of an immature bird which was taken in Cape York Peninsula in 1866 also had a red suffused face. I have found that wild immature birds can be readily sexed by the brightness of the red face and ear coverts. Within three or four months of leaving the nest the young birds attain adult plumage. The interesting feature of the red suffused face in immature wild Varieds is that I have never bred, seen, nor can find records of an aviary-bred bird having a red face.

Aviculture

The first European aviculatural record of this species appears to be the importation into London of several pairs in 1902. In 1903 a Varied Lorikeet was exhibited at the Crystal Palace Bird Show in London. It was entered in the parakeet class and it won first prize. The first recorded breeding was by E. J. Boosey at the Keston Bird Farm in England in 1936. In Australia, C. J. (Charlie) Lambert, a well-known aviculturist of his day, from the outer Sydney suburb of Horsley, recorded the first breeding in 1949. Eggs were laid on the 16th and the 18th of September; incubation period was 22 days and the chicks fledged in 37 and 39 days respectively and were independent in three weeks. The diet consisted of Weetbix, milk, sugar, honey, apple and milk thistle. One chick left the nest with a black beak and the other with a pinkish beak.

Like all other small lorikeets, this species is bred far more regularly now than in the past. This is mainly due to the diets and housing techniques being used by Australian aviculturists at present, but I still feel this is the most difficult of the Australian lorikeet species.

Flock breeding has been successful in this species. My first successes were with a flock of three pairs housed in a planted aviary approximately eight metres square. However, I think best results are obtained with single pairs housed in smaller aviaries or suspended wire cages. I have found Varieds to be more delicate than the other species and their housing should be well protected and sheltered from the prevailing winds.

Diet

Particular care should be taken with the diet of this species as they are quite prone to obesity and they become very lethargic.

Sexing

Unlike any of the other Australian lorikeets, this species is sexually dimorphic (i.e. the sexes differ in appearance from each other). The female is generally duller, the red cap is duller and smaller, the yellow ear coverts and the yellow striations are not as bright and the mauve pink of the breast is far less extensive than in the male. When one has difficulty sexing Varieds, it is because all the birds are of the one sex.

Display

Courtship display is similar to other lorikeets with the male drawing himself to his full height, dilating his pupils, arching his neck and bobbing while hopping along the perch. However, this species does not display as readily or as flamboyantly as other lorikeet species.

Nest Boxes

The nest box I have found to be the most preferred is as suggested for the Little Lorikeet. See Fig. 4. Some pairs will frequent and sleep in their nest box, but not to the extent of the Little or Purple-crowned Lorikeets. In fact, many pairs only enter the nest box when breeding is in progress.

Nesting and Hatching

I have recorded three to five eggs laid at two- or three-day intervals; incubation usually starts with the second egg and extends for a duration of 20-23 days, with the majority hatching in 22 days.

The chicks hatch with a whispy pale grey down, their eyes start to open and the secondary down begins to appear around 10 days of age and pin feathers start to develop soon after. Chicks are fully feathered at 35 days and fledging has occurred in my aviaries from 37 to 41 days of age, with the average being 39 days. This is five days shorter than the Little Lorikeet and 13 days less than the Purple-crowned Lorikeet.

All aviary bred Varieds which I have observed, both in my own aviaries and in other peoples' aviaries, have left the nest similar to the adult female, without a red cap, but with a reddish orange frontal band (that area of the brow above the beak and between the eyes). The beak is brown tinged with orange. The immatures are easily sexed with the males being generally brighter, particularly on the frontal band and ear coverts. There is no sign of the reddish suffusion around the facial areas as seen in immature wild birds. Within three months of leaving the nest, the red cap has moulted in, the beak is a yellow orange, and the youngsters are duller replicas of their parents. At 12 months of age they are fully adult and capable of breeding.

Breeding Season

The usual breeding season in the Sydney region is mid-June until early October. Peter Hobbs, from Gloucester, NSW, had a pair nest in late November, but he assures me this was unusual for his locality. While attending the National Avicultural Convention in Melbourne in 1985, Barbara O'Brien showed me her Varied Lorikeets on eggs in early May and did not consider this to be unusual for that area. I felt that it was quite unusual as most Victorian Aviculturists consider their birds breed later than the aviary birds of NSW. Likewise, the NSW breeding season is considered to be later than the Queensland breeding season.

In my own aviaries in the south-western suburbs of Sydney, they have always been single brooded if left to rear their own young, and have nested from July to September.

Mutations

I can find no record of mutations occurring in this species. A fertile hybrid has been recorded with the Red-collared Lorikeet, which in turn produced a hybrid with a Rainbow Lorikeet. The offspring, when mated with a Scaly-breasted Lorikeet, again produced fertile eggs (Noel Dunstan, personal communication).

MUSK LORIKEET

(Glossopsitta concinna)

PLATE 16: Musk Lorikeet (*Glossopsitta concinna*)

MUSK LORIKEET

(Glossopsitta concinna)

Derivation
Glosso — Greek for tongue
psitte — Greek for parrot
concinnus — Latin for pretty

Description
Length 22 cm
Weight 59 grms average
Appearance as in Plate 16

Classification

I feel that the Musk Lorikeet does not fit comfortably into this genus. It has a distinctive slender black upper mandible with a bright orange tip which is unlike that of any other lorikeet. In addition it is a large bird and lays a clutch of two eggs compared to three to five eggs laid by the other members of the genus. Perhaps these differences are not enough to suggest it be installed into a monotypic genus (a genus having only one member), but it is certainly dissimilar to the other Australian lorikeets.

Earliest Reports

Like the Rainbow Lorikeet, the Musk Lorikeet featured in our early history, being described and figured by Captain Phillip in his "Voyage to Botany Bay" 1789.

Range and Habitat

The range extends through eastern and south-eastern Australia from the Bowen district in north Queensland through NSW and Victoria, to the Spencer Gulf region of South Australia, and to the eastern half of Tasmania. It is less abundant in the northern parts of its range.

This species is nomadic, but more seasonal in its movements than other lorikeets. I can find no evidence of resident populations being established as with *Trichoglossus*. Numerous items published in the RAOU's "The Emu", since its inception in 1900, provide evidence of the seasonal nature of this bird's movements. There is no colour or size variation throughout its range, which is indicative of constant population integration.

I have usually seen this species in pairs or small scattered flocks in areas where eucalypts are flowering profusely, but at times they congregate in large flocks when conditions are favourable or when usual feeding grounds are limited through bushfires or snow falls, etc. They are found in most types of habitat within their range from coastal scrubs to open forest and to the dried mallee country; wherever flowering or fruiting trees occur. They are uncommon in mountainous and high-altitude country. Their liking for the sweet waxy scale-like secretion of the sugar lerp insect has been noted by many observers. When Musk Lorikeets are seen foraging through the new leaf growth on many of our native trees, particularly eucalypts, they are most likely to be feeding on lerp secretion. Crop damage in soft fruit orchards is common.

The Musk Lorikeet is often seen feeding in the company of Rainbow, Scaly-breasted, Little and Purple-crowned Lorikeets and the Swift Parrot.

Breeding in the Wild

Nesting usually takes place high up in hollow limbs or holes in trees, most often in eucalypt trees. The nest site is prepared by both male and female and is usually about 0.5 metre deep on a base of chewed-up decaying wood. The usual clutch is two white eggs which soon become soiled in the nest. The only record I can find of more than two eggs in a clutch is a set of four eggs in the H. L. White collection in the Australian Museum, Melbourne, which I feel must be regarded as doubtful.

The breeding season extends from late winter until midsummer. Birds in the northern parts of the range usually breed early in the season, and later in the southern parts.

The name Musk Lorikeet was given to this species due supposedly to its musky odour. I have never been able to detect this odour in any of my birds or in any of their nests of babies being raised. I have, however, noticed quite unpleasant sickly odours coming from cages of obviously trapped birds on dealers' premises in years gone by. This odour, I feel, was more related to overcrowding, excrement and the sickly honey based slop being fed to the birds. On one occasion I can remember an unusual sweet odour in a group of wild caught Musks. The smell was the same as often encountered in wild caught Australian parrakeets, particularly the freshly imported Western Australian Red-capped Parrakeet (*Purpureicephalus spurius*).

Aviculture

The first recorded breeding of this species is credited to Herr Negel of Germany in 1903, with a pair housed in a birdroom. The pair bred in a nest box hung high in the birdroom. During the rearing period the female died and the male continued to feed the chicks and he raised them successfully. This breeding took place during the late winter and the early spring.

Most literature relating to the keeping of Musk Lorikeets in aviaries suggests that they are difficult to keep in captivity, subject to obesity and even that they are excessively aggressive in mixed company. I cannot agree with any of these statements. When fed a sensible and well balanced diet, they are a hardy, active and delightful aviary subject, even when housed in suspended cages. I have housed and bred this species in large mixed aviaries containing doves, finches, neophemas, princess parrots and Purple-crowned Lorikeets, without their showing the slightest sign of aggression, even when rearing young. Of course, I will admit that every now and then in all species of *psittacines* a rogue will occur, but I can assure you that in Musk Lorikeets aggressive behaviour is the exception, not the rule.

Flock breeding can be successful in spacious surroundings but best results are obtained one pair per aviary.

Sexing

Sexual differences can be quite noticeable in some pairs, with the male showing more and brighter blue on the crown, but this is by no way a general means of sexual identification. I have seen some extremely bright blue crowned females and some very dull males. Females often appear to be larger than males. Surgical sexing is always advisable when pairing this species.

Display

The courting display is far less demonstrative than in the *Trichoglossus* species and is usually limited to some bobbing up and down with the neck arched and the pupils dilating and sometimes accompanied with a low whistling note. The pair bond, as with most of the lorikeets, is strong and pairs can often be seen sitting close together engaged in mutual preening.

Nest Boxes

As with the Scaly-breasted Lorikeet, for convenience, I use a nest box approximately 23 cm by 23 cm by 30 cm deep with a 6 cm or 7 cm diameter entrance hole and perch located near the top of the box. See Fig. 5. However, a box with a larger horizontal area could be advisable.

Nesting and Hatching

Two white eggs are laid at two- or three-day intervals. Incubation is carried out by the female and most males sleep in the nest box with the incubating female. My records show a greater variation in the incubation period for this species than for any other. That is from 19 days to 25 days with the average incubation period being 22 days. I cannot give any explanation for this extraordinary variation in incubation period, not even weather conditions, as the 19-day period and the 25-day period both took place under controlled conditions in October and November, 1983, by two separate pairs. Rosemary Low, 1977, also records a large variation in the incubation

A

B

C

D

PLATE 17: Creating a New Colour.
A. Olive hybrid, 1/2 Olive Scaly-breasted, 1/2 Musk.
B. Olive hybrid, 3/4 Musk, 1/4 Scaly-breasted.
C. Olive hybrid, 7/8 Musk, 1/8 Scaly-breasted.
D. Young 15/16 Musk Lorikeets.

PLATE 18: Musk Lorikeet Nestlings at 2 days, 18 days, 4 weeks and 6 weeks old respectively.

period for Musk Lorikeets.In 1986 Jim Gill recorded a 26-day and a 27-day incubation period from a pair that had previously had incubation periods of 23 days.

This species nests seasonally in my aviaries with the earliest egg recorded being laid on 5th July and the last egg of a season being laid on 25th November.

Like the *Trichoglossus*, Musks hatch with a long silvery white down and start to develop grey secondary down about 12 days of age. They open their eyes at around 14 days of age and the commencement of pin feather development occurs a few days later.

The average fledging period in my aviaries was 48 days. The young leave the nest a duller version of their parents, with orange-red facial markings, a dark brown beak and eye, and the cere and the bare skin around the eye is pale grey. By three months of age the beak tip has acquired the orange colouring and the eye colour is similar but duller than that of the adult. The red markings of the head often do not reach the intensity of the adults until six months old. I have bred from 12 months old birds, but I have found this to be the exception, with most birds not breeding until their second year.

Fertile hybrids have been recorded with the Rainbow, Red-collared, Scaly-breasted and Little Lorikeets.

Mutations

I can find no record of natural mutations occurring in this species. Birds with mottled yellow areas which are sometimes quite extensive have been recorded both in the wild and in aviaries. This is an acquired variation, and in all the cases that I have been familiar with the yellow has either moulted out or the bird has died. This suggests that the disorder is due to a nutritional deficiency or to a systemic illness in the bird.

Through hybridising the Musk Lorikeet with the Olive Scaly-breasted Lorikeet, after four generations I have produced Olive Musk Lorikeets. The birds retain all red and yellow colouring, the green becomes olive green and the blue becomes grey. See Plate 17.

LITTLE LORIKEET

(Glossopsitta pusilla)

PLATE 19: Little Lorikeet (*Glossopsitta pusilla*).

LITTLE LORIKEET

(Glossopsitta pusilla)

Derivation
Glosso — Greek for tongue
psitte — Greek for parrot
pusilla — Latin for very small

Description
Length 15 cm
Weight 44 gms
Appearance as in Plate 19

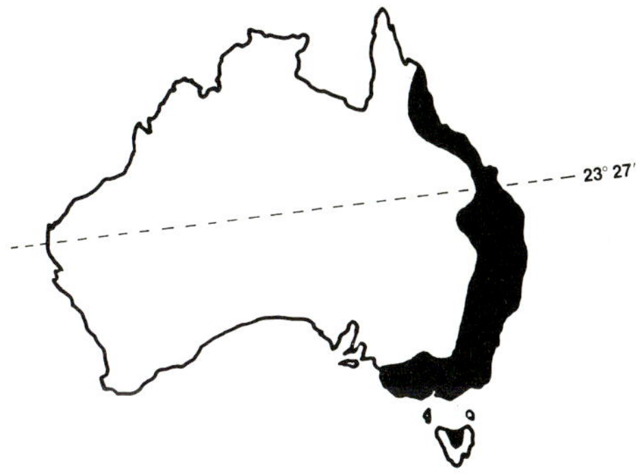

Classification

I consider this lorikeet to be the typical member of this totally Australian genus. There have been no subspecies recorded in this extremely nomadic lorikeet, due to the integration throughout most of the range, keeping the species constant in colour and form.

Earliest Reports

Like many of our native parrots, the Little Lorikeet did not escape the attention of the early naturalists. White figured and described this species in his "Journal of a Voyage to NSW" in 1789.

Range and Habitat

This species ranges through eastern and south-eastern Australia, from the Cairns district of North Queensland to York Peninsula in South Australia and occasionally into north-eastern Tasmania. It is strictly nomadic occurring only where there is a suitable food supply, particularly flowering eucalyptus and Melaleuca trees. I have mostly observed this species in small parties of up to 10 or so birds, but on occasions flocks of a hundred or two may be seen. When congregating in these large flocks, they fly swiftly and noisely in unison from flowering tree to flowering tree, all arriving in a tree at breakneck speed to feed and then all simultaneously departing amid much screeching. When these birds first arrive in an area they feed on the blossom at the top of the trees, but as the higher blossom is eaten out over a period of weeks, they gradually feed lower and lower until quite close observation can be made of them clambering about the lower blossom. This species is often seen feeding in the company of Rainbow, Scaly-breasted and Musk Lorikeets and in Victoria and South Australia with the Purple-crowned Lorikeet.

Little Lorikeets have been observed feeding on new leaf tips and gleaning leaves and branches for insect life. They have been known to cause crop damage in some soft fruit orchards but pollen and some nectar form their basic diet.

Breeding in the Wild

This species nests in hollow limbs, or holes in trees. They usually prefer a nesting site 6 to 12 metres above the ground, with the entrance via a very small knot hole. Some nests have been found to be soiled with excrement around the entrance hole and quite smelly inside when young are being reared. Other nests in different localities have been quite clean and dry. Perhaps the diet of the parent bird has some bearing on this.

Three to five white rounded eggs are laid in the hollow which is prepared by both the male and female. Nesting usually takes place in late winter and spring, but has been recorded as early as May in North Queensland and as late as January in the southern extremity of its range.

The sexes are virtually identical, but some older males have a larger, brighter red mask.

As a child, this was the only species of lorikeet I was familiar with, because it was the only one to occur with any regularity in the south-western suburbs of Sydney where I was raised. I can remember gazing at these little "keets" as we called them, feeding on the blossom of the iron bark eucalypts adjacent to our home, pondering how I could capture some. It was to be 30 years before I had the opportunity to keep and breed this species. My home now stands where those iron bark trees once stood, but we still have an occasional visit from Little Lorikeets. They are not in the numbers of days gone by, just odd pairs that drop out of the sky when they hear my aviary birds calling. They usually land in adjacent trees, make a few low passes over the aviaries and then fly off; odd ones may linger for an hour or two, calling excitedly to my birds.

Aviculture

The Little Lorikeet appears to have been imported into Europe as early as 1877, but the lack of early breeding records or, indeed, any avicultural information, reflects the problems the pioneering aviculturists had with this species. The first breeding success is credited to N. K. Bush of Peakhurst, Sydney, NSW, in 1948. His pair were housed in a spacious planted aviary together with a mixed finch collection. The chosen nest box was hung in the sheltered area of the aviary and was fitted with a natural hollow which acted as an entrance spout. The male was a hand raised bird and the female was wild caught. A clutch of four eggs was laid in September, hatched in mid-October and fledged in November, and the young were independent of the parents within two weeks. The Bush family have been prominent in Sydney avicultural circles for well over 50 years and younger members of the family are still avid aviculturists.

Diets

Over the last few years, with the development of diets more suitable to the specialised requirements of this species, successful breedings are occurring regularly and there are now well established aviary bred strains.

Sexing

Successful flock breeding in large aviaries has been recorded, but all my successes have been with single pairs housed in wire bottomed suspended cages. Surgical sexing is advisable, for although some older males can be distinguished from the female by their larger and deeper red mask and the darker brown colouring of the nape and mantle, this is by no means reliable. Some males I have bred from are quite dull and indistinguishable from the female.

Display

The courting male approaches the female stretched to his full height, but without the arching neck as seen in the *Trichoglossus*. During his advances he bobs and hops and wipes his beak frequently along the perch as he advances towards the hen. The dilating of the pupil of the eye is not as noticeable as in the larger species. During the display the male often emits a soft, low whistling note.

Nest Boxes

I have found the nest boxes that I use for African Lovebirds to be the most suitable for Little Lorikeets. This is a horizontal box 30 cm long, 15 cm wide and 17.5 cm deep, with a small entrance hole at one end and near the top of the large side. See Fig. 4. Small holes 3 mm in diameter drilled through the bottom of the box can help to keep the nest dry. The box should be positioned at approximately 15 degrees from the horizontal to keep the eggs together at the end of the box. Pairs usually sleep in the box even when not breeding so it is advisable to make nest boxes available to all birds whether in breeding or holding facilities.

Nesting and Hatching

My records show that three to five rounded white eggs (average 4) are laid at two- to three-day intervals (average two days). I have found that nesting is seasonal in my aviaries, the earliest egg recorded being laid on 26th June and the latest egg to be laid in any season was on 21st October. Of course, this breeding period would vary with climatic conditions. I would

A B C D

PLATE 20: Little Lorikeet Nestlings at 1 day, 2 weeks, 4 weeks and 5 weeks old respectively.

expect that birds kept in milder conditions nearer the coast in the Sydney area would have a longer breeding period. Some pairs are double brooded.

The incubation period ranged from 20 days to 22 days with the vast majority of chicks hatching in 20 days. Chicks hatch with a silvery white down which is replaced gradually with a shorter, thicker grey secondary down from 10 days of age. The eyes start to open at the same time. Pin feathers are developing by 20 days and full feather development is attained between 35 and 40 days. Fledging has occurred between 39 and 46 days, with the average time spent in the nest in my aviaries being 44 days.

When the chicks leave the nest they are duller replicas of the adults, with dark brown beak and iris. The cere and the bare skin around the eyes are noticeably paler grey. They can be removed from the parents three weeks after leaving the nest. Within three months of leaving the nest the only variation from the adults is that they are slightly duller and smaller. They are fully adult and capable of breeding at 12 months of age.

Mutations

I have been told that a lutino exists in a north Queensland aviary, but as yet this is unconfirmed. Currently I am trying to hybridise an olive Musk Lorikeet with a Little Lorikeet. If this project is successful, perhaps an olive Little Lorikeet will eventually be produced. Fertile hybrids with the Musk Lorikeet have been recorded.

PURPLE-CROWNED LORIKEET

(Glossopitta porphyrocephala)

PLATE 21: Purple-crowned Lorikeets (*Glossopsitta porphyrocephala*). In this pair the female is the brighter bird.

PURPLE-CROWNED LORIKEET

(Glossopitta porphyrocephala)

Derivation
Glosso — Greek for tongue
psitte — Greek for parrot
porphyrois — Greek for purple
cephale — Greek for head

Description
Length 16 cm (7")
Weight 45 gm average
Appearance as in Plate 21

Classification

A typical member of the *Glossopsitta* genus which shows no variation through the extensive east-west range, except for some individual colour variants, in which the frontal band may range from yellow to red. Subspecies have never been suggested.

Earliest Reports.

The Purple-crowned Lorikeet escaped the attention of the early naturalists and was named by Dietrichsen in 1832. The first written account and illustration of this species appears to be Gould's "Birds of Australia", 1848.

Range and Habitat

It ranges across southern Australia including all the south-west, extending north almost to Shark Bay, Western Australia. In the south-east it extends north to Eden, NSW. The range is broken by the barren area of the Nullarbor Plain at the northern extremities of the Great Australian Bight. Whether or not this natural barrier has divided this species into two separate populations can only be supposition. The recorded sightings in northern New South Wales and southern Queensland suggest that at times this species will travel great distances from its normal range, so it is quite possible that the eastern and western populations do intermingle. The absence of this bird from Tasmania, and the fact that it is the only lorikeet native to south Western Australia, is thought to suggest that it is a south-western species that has extended its range east, but has not yet reached Tasmania.

Being a nomadic species, its numbers may vary dramatically in a given area from season to season and from year to year. In some localities, particu larly the south-west, small resident populations do appear to occur. The habitat varies from the tall karri and jarrah forests of the south-west, coastal forests and scrubs and to the dry mallee country. It feeds on pollen, nectar, the fruits of native trees and shrubs and insects. At times it also feeds on cultivated fruits, causing considerable crop damage. This lorikeet is often seen feeding in the company of Little and Musk Lorikeets.

I have only seen this species in small parties, but at times dense flocking will occur. This may be brought about by profusely flowering trees in a particular area or adverse conditions in a neighbouring area normally frequented by them. Such a concentration occurred in Port Pirie, South Australia, a few years ago, and was related to me by an old friend and master aviculturist, the late Ross Hogben of that city. Ross told me the Purple-crowns arrived in thousands and could be seen flying and screeching everywhere. They were hanging from the blossoms of every flowering tree and shrub throughout the city. Children were plucking them from low blossoms, where some appeared to be intoxicated or perhaps weakened by hunger due to the large concentration sharing the food supply. Many injured birds were brought to Ross after flying into windows or wires. Unfortunately, many more were caught after going into poultry sheds and other enclosures in search of water. Many people who caught these birds thought they would make a pretty pet but did not have the slightest idea of their dietary requirements. Then, as suddenly as they arrived, they departed.

Breeding in the Wild

Breeding usually takes place in spring and summer from August to December. Although scattered pair breeding is usual, large breeding colonies have been recorded, with many pairs nesting in close vicinity to each other. Often two or more pairs have been seen occupying hollows in the same tree. The nesting site is a hollow limb or hole in a tree, often in eucalypt, with a preference shown for a small entrance hole. Nesting sites have been recorded as low as three metres and as high as 40 metres from the ground. The nest hollow is prepared by both the male and the female. Three or four white eggs are laid, usually on a bed of chewed up decaying wood. Incubation is carried out by the female with the male sleeping in the hollow at night.

Aviculture

The first importation into Europe appears to have been in 1910, but the lack of early avicultural information on this species reflects the difficulties European and English aviculturists had with this lorikeet. This bird is said to be totally absent from European and English aviaries today.

The first successful breeding appears to have been in San Diego Zoo, USA, in 1930, although precise records are not avilable. In Australia the first recorded breeding was by J. Gregg of Croydon, NSW, in 1936, where two young were reared. Charlie Lambert, that successful aviculturist of the 1940's and 50's from Horsley in Sydney's west, bred this species in 1952. The pair were fed on a porridge made from Farex, crushed biscuit, plain cake, honey, milk, sugar and chopped apple.

Diets

Like the Little Lorikeet this species has responded well to modern lorikeet diets and techniques. I consider it to be the most prolific and successful of the small Australian lorikeets in our aviaries. In the Sydney area it will breed at any time of the year and my records show that eggs have been laid in my aviaries in every month of the year except January and February which are our two hottest months. Although young in the nest hatched from eggs laid in December survived the heat of these months.

Flock Breeding

Many aviculturists have had good results from flock breeding in large planted aviaries. Kath Thompson of South Australia and Lynne Williams of Perth, Western Australia, have both succeeded in this area. I have bred these birds under flock conditions but my best results by far have been from single pairs, housed in wire-bottomed suspended cages. Surgical sexing is definitely advisable when selecting pairs in this species.

On occasions a breeding pair, housed in a wire cage for a couple of years, has become overweight and stopped breeding. A few months flying in a large aviary trims them down and returns them to breeding condition.

Sexing

I have found that males and females are similar but seldom totally alike. There are slight differences such as variation in the colour and size of the frontal band, cheek patches and purple crowned area. This differs from bird to bird and is not an indication of sex. At times I feel that I can detect differences in the size and shape of the head, with the male having a slightly rounder, larger head and the female being flatter on the crown.

Display

The male is not as demonstrative as the Little Lorikeet. His advances usually consist of stretching himself to his full height and bobbing his head while hopping along the perch towards the female. Again the dilation of the pupil is not as noticeable as in the *Trichoglossus*, perhaps because of the darker eye colour in this species.

Nest Boxes

The nest box used is the same as for the Little Lorikeets. See Fig. 4. Pairs will frequent the nest box during the day and sleep in it all year round. They will often fly into the box if startled or approached by strangers.

Nesting and Hatching

Three or four eggs are the usual clutch laid in my aviaries, more than four never having been recorded. The recorded incubation periods vary from 19 to 22 days, with the most common period being 20 days. Incubation is carried out by the female, although the male spends much of his time in the nest with the hen. Chicks hatch with a silvery grey down, which, commencing at 12 days, is gradually replaced with a shorter grey secondary down and the eyes begin to open at this same period. Pin feather development is noticeable at 22 days and the chicks are feathered by 45 days.

Fledging has taken place in my aviaries between 45 and 60 days with the most usual period being around 52 days. There has been a considerable variation of the fledging period for Purple-crowns in my aviaries, with the minimum time spent in the nest being equal to the maximum time recorded for Little Lorikeets. I find it hard to explain the variation in fledging periods in birds from the same genus that range into the same areas and feed on identical diets. All my data on these species was taken during the same period of time with both species being housed and fed identically.

When chicks leave the nest they are considerably duller in colour than the adults, with no traces of red on the facial area and with the purple crown reduced, if not absent. The cere and bare skin around the eye is a paler grey and the iris is a dark brown. Full colour is attained about three months after leaving the nest and they are sexually mature at six months of age. A six-month-old pair has reared young in my aviaries.

I have found Purple-crowns to be safe when housed in a large aviary with finches, doves and larger parrots, although one pair in my collection showed aggression towards Neophema Parrots.

Mutations

The only record I can find of a mutation in this species is that recorded in the "Emu", Volume 59, page 83, by E. J. Boehna of Sutherland, South Australia. He states that on the Mount Mary Plains, South Australia, in October, 1927, a nest was found containing two albino young. He describes them as a lovely clear bright yellow with cream remiges and rectrices with the red, orange and purple colours unchanged and with pink eyes. From this description I feel they could not have been lutinos or albinos as the purple would have had to change to white. The mutation described must be an advanced cinnamon yellow and I shall record it as such.

B

D

PLATE 22: Purple-crowned Lorikeet Nestlings at 1 day, 3 weeks, 5 weeks and 7 weeks old respectively.

The prominent and successful Sydney aviculturist, Sid Gale, inadvertently bred a hybrid between a Purple-crowned Lorikeet and a Little Lorikeet in a large planted aviary in the early 1980's. Unfortunately, the bird was disposed of without being photographed or its colouring described.

REFERENCES

Forshaw, J., Australian Parrots, Lansdowne Press.

Harrison and Harrison, Clinical Avian Medicine and Surgery, Saunders.

Low, R., Lories and Lorikeets, Paul Elek Ltd.